SCIENTIFIC AMERICAN

Current Issues in Biology, Vol. 5, is published by Scientific American, Inc. with project management by:

DIRECTOR, ANCILLARY PRODUCTS: Diane McGarvey
CUSTOM PUBLISHING MANAGER: Marc Richards
CUSTOM PUBLISHING EDITOR: Lisa Pallatroni
DESIGNER, FRONT MATTER: Silvia De Santis

Instructors:

For correlations of articles to the chapters in your Benjamin Cummings biology titles, as well as accompanying PowerPoint presentations and suggested answers to the "Biology in Society" and "Thinking About Science" review questions, please visit the companion website to *Current Issues*, Volume 5 at
http://www.aw-bc.com/info/scientificamerican/volume5.html

The contents of this issue are adaptations of material previously published in SCIENTIFIC AMERICAN.

EDITOR IN CHIEF: John Rennie
EXECUTIVE EDITOR: Mariette DiChristina
MANAGING EDITOR: Ricki L. Rusting
NEWS EDITOR: Philip M. Yam
SENIOR WRITER: Gary Stix
SENIOR EDITOR: Michelle Press
EDITORS: Mark Alpert, Steven Ashley, Graham P. Collins, Mark Fischetti, Steve Mirsky, George Musser, Christine Soares

ART DIRECTOR: Edward Bell
ASSOCIATE ART DIRECTOR: Mark Clemens
ASSISTANT ART DIRECTOR: Johnny Johnson
PHOTOGRAPHY EDITOR: Emily Harrison
PRODUCTION EDITOR: Richard Hunt

COPY DIRECTOR: Maria-Christina Keller
COPY CHIEF: Daniel C. Schlenoff
COPY AND RESEARCH: Michael Battaglia, Smitha Alampur, Michelle Wright, John Matson, Aaron Shattuck

EDITORIAL ADMINISTRATOR: Jacob Lasky
SENIOR SECRETARY: Maya Harty

ASSOCIATE PUBLISHER, PRODUCTION: William Sherman
PREPRESS AND QUALITY MANAGER: Silvia De Santis
PRODUCTION MANAGER: Christina Hippeli

ASSOCIATE PUBLISHER, CIRCULATION: Simon Aronin

ASSOCIATE PUBLISHER, STRATEGIC PLANNING: Laura Salant

GENERAL MANAGER: Michael Florek
BUSINESS MANAGER: Marie Maher

CHAIRMAN: Brian Napack

PRESIDENT: Steven Yee

VICE PRESIDENT AND PUBLISHER: Bruce Brandfon

VICE PRESIDENT: Frances Newburg

ON THE COVER
top: Kenn Brown;
bottom left: James Porto
bottom right: Scott Gibson

Current Issues in Biology, volume 5, published by Scientific American, Inc., 415 Madison Avenue, New York, NY 10017-1111. Copyright © 2007 by Scientific American, Inc. All rights reserved. No part of this issue may be reproduced by any mechanical, photographic or electronic process, or in the form of a phonographic recording, nor may it be stored in a retrieval system, transmitted or otherwise copied for public or private use without written permission of the publisher.

Subscription inquiries for SCIENTIFIC AMERICAN magazine:
U.S. and Canada (800) 333-1199; other (515) 247-7631, or www.sciam.com.

To learn more about Scientific American's Custom Publishing Program, contact Marc Richards at 212-451-8859 or mrichards@sciam.com.

CANCER Clues from

Studies of pet dogs with cancer can offer unique help in the fight against human malignancies while also improving care for man's best friend

Imagine a 60-year-old man recuperating at home after prostate cancer surgery, drawing comfort from the aged golden retriever beside him. This man might know that a few years ago the director of the National Cancer Institute issued a challenge to cancer researchers, urging them to find ways to "eliminate the suffering and death caused by cancer by 2015." What he probably does not realize, though, is that the pet at his side could be an important player in that effort.

Reaching the ambitious Cancer 2015 goal will require the application of everything in investigators' tool kits, including an openness to new ideas. Despite an unprecedented surge in researchers' understanding of what cancer cells can do, the translation of this knowledge into saving lives has been unacceptably slow. Investigators have discovered many drugs that cure artificially induced cancers in rodents, but when the substances move into human trials, they usually have rough sledding. The rodent models called on to mimic human cancers are just not

PET DOGS

By David J. Waters and Kathleen Wildasin

DOGS AND HUMANS often fall ill with the same kinds of cancers. Scientists contend that the similarities between these tumors, including genetic resemblances, can be instructive. (The background represents the DNA sequence from a tissue sample.)

measuring up. If we are going to beat cancer, we need a new path to progress.

Now consider these facts. More than a third of American households include dogs, and scientists estimate that some four million of these animals will be diagnosed with cancer this year. Pet dogs and humans are the only two species that naturally develop lethal prostate cancers. The type of breast cancer that affects pet dogs spreads preferentially to bones—just as it does in women. And the most frequent bone cancer of pet dogs, osteosarcoma, is the same cancer that strikes teenagers.

Researchers in the emerging field of comparative oncology believe such similarities offer a novel approach for combating the cancer problem. These investigators compare naturally occurring cancers in animals and people—exploring their striking resemblances as well as their notable differences.

Right now comparative oncologists are enlisting pet dogs to tackle the very obstacles that stand in the way of achieving the Cancer 2015 goal. Among the issues on their minds are finding better treatments, deciding which doses of medicines will work best, identifying environmental factors that trigger cancer development, understanding why some individuals are resistant to malignancies and figuring out how to prevent cancer. As the Cancer 2015 clock keeps ticking, comparative oncologists ask, Why not transform the cancer toll in pet dogs from something that is only a sorrow today into a national resource, both for helping other pets and for aiding people?

Why Rover?

FOR DECADES, scientists have tested the toxicity of new cancer agents on laboratory beagles before studying the compounds in humans. Comparative oncologists have good reason to think that pet dogs with naturally occurring cancers can likewise become good models for testing the antitumor punch delivered by promising treatments.

One reason has to do with the way human trials are conducted. Because of the need to ensure that the potential benefits of an experimental therapy outweigh the risks, researchers end up evaluating drugs with the deck stacked against success; they attempt to thrash bulky, advanced cancers that have failed previous treatment with other agents. In contrast, comparative oncologists can test new treatment ideas against early-stage cancers—delivering the drugs just as they would ultimately be used in people. When experimental drugs prove helpful in pets, researchers gain a leg up on knowing which therapies are most likely to aid human patients. So comparative oncologists are optimistic that their findings in dogs will be more predictive than rodent studies have been and will help expeditiously identify those agents that should (and should not) be tested in large-scale human trials.

Pet dogs can reveal much about human cancers in part because of the animals' tendency to become afflicted with the same types of malignancies that affect people. Examples abound. The most frequently diagnosed form of lymphoma affecting dogs mimics the medium- and high-grade B cell non-Hodgkin's lymphomas in people. Osteosarcoma, the most common bone cancer of large- and giant-breed dogs, closely resembles the osteosarcoma in teenagers in its skeletal location and aggressiveness. Under a microscope, cancer cells from a teenager with osteosarcoma are indistinguishable from a golden retriever's bone cancer cells. Bladder cancer, melanoma and mouth cancer are other examples plaguing both dog and master. In a different kind of similarity, female dogs spayed before puberty are less prone to breast cancer than are their nonspayed counterparts, much as women who have their ovaries removed, who begin to menstruate late or who go into menopause early have a reduced risk for breast cancer.

Canine cancers also mimic those of humans in another attribute—metastasis, the often life-threatening spread of cancer cells to distant sites throughout the body. Solving the mystery of how tumor cells metastasize to particular organs is a top research priority. When certain types of cancers spread to distant organs, they tend to go preferentially to some tissues over others, for reasons that are not entirely clear. Because metastasis is what accounts for most deaths from cancer, researchers would very much like to gain a better understanding of its controls. Studies in pet dogs with prostate or breast cancer might prove particularly useful in this effort, because such tumors frequently spread in dogs as they do in humans—to the skeleton. Indeed, research in pet dogs is already attempting to work out the interactions between tumor cells and bone that make the skeleton such a favorite site for colonization.

Scientists also have deeper theoretical grounds for thinking that pet dogs are reasonable models for human cancer. Evolutionary biologists note that dogs and humans are built like Indy race cars, with successful reproduction as the finish line. We are designed to win the race, but afterward it does

Overview/Canine Cancer

- Millions of pet dogs will have cancer diagnosed this year. In many of those animals, the malignancy will look and behave much as it would in humans, such as spreading to the same organs.
- Investigation of these cancers can help researchers to better understand the biology of the human forms. Also, studies of experimental treatments in the animals can indicate which therapies most deserve further testing in dogs and humans and can offer guidance on the best doses and methods of delivery.
- Such studies should improve cancer prevention and therapy for both humans and their canine companions.

BREEDS AT RISK

The breeds represented by the dogs shown here are particularly susceptible to cancers that also afflict humans. These malignancies look like the human forms under a microscope and act similarly as well. Such resemblances mean that canine responses to experimental drugs should offer a good indication of how the compounds will work in humans. In addition, research into the genes that increase susceptibility of specific breeds to particular cancers is expected to help pinpoint susceptibility genes in humans.

Rottweiler: Bone cancer

Collie: Nasal cancer

Chow Chow: Stomach cancer

Boxer: Brain cancer

Golden Retriever: Lymphoma

Scottish Terrier: Bladder cancer

SKELETAL DISTRIBUTION of metastases is another aspect of cancer similar in dogs and humans. In dogs, the lesions display the same "above the elbow, above the knee" pattern seen in people. Insights into why that pattern occurs in dogs could help explain the distribution in humans and perhaps suggest new ideas for intervening. (The numerals indicate the number of metastases found at each site in one study.)

Maxilla, 1
Scapula, 4
Humerus, 29
Radius, 3
Ulna, 7
Metacarpal, 1
Vertebra, 42
Pelvis, 7
Femur, 24
Rib, 32
Sternum, 3
Tibia, 14

not matter how rapidly we fall apart. This design makes us ill equipped to resist or repair the genetic damage that accumulates in our bodies. Eventually this damage can derange cells enough to result in cancer. In the distant past, our human ancestors did not routinely live long enough to become afflicted with age-related cancers. But modern sanitation and medicine have rendered both longevity and cancer in old age common. Much the same is true for our pets. Pet dogs, whom we carefully protect from predation and disease, live longer than their wild ancestors did and so become prone to cancer in their later years. Thus, when it comes to a high lifetime risk for cancer, pets and people are very much in the same boat.

Aside from acquiring cancers that resemble those in people, pet dogs are valuable informants for other reasons. Compared with humans, they have compressed life spans, so scientists can more quickly determine whether a new prevention strategy or therapy has a good chance of improving human survival rates. Finally, although veterinarians today are far better equipped to treat cancer than they used to be, the standard treatments for many canine tumors remain ineffective. Because most pet cancer diagnoses end in death, dog owners are often eager to enroll their animals in clinical trials that could save their pet's life—and possibly provide the necessary evidence to move a promising therapy to human clinical trials.

The Ideal Animal Model: An Invalid Concept

RODENTS are a favorite model of cancer researchers, but therapies that work beautifully in rats and mice often fail in humans.

Some experts contend that progress toward finding cancer cures has been frustratingly slow because of the inadequacy of available animal models of human cancer. But perhaps the problem is not in the animals themselves but in the way they are used and what we are forcing them to tell us.

The dictionary defines a model as "an imitation." By definition, therefore, an animal model of cancer is not the same as a person who acquires cancer. Rodent models are often produced by making "instant cancers"—that is, by injecting the animals with tumor cells or bombarding them with carcinogen doses that are higher than any human will ever encounter. It is doubtful that cancers produced in that way will accurately recapitulate a complex process that often requires more than 20 to 30 years to develop in people. Naturally occurring animal tumors, such as those affecting pet dogs, provide the opportunity to study this complexity in a less artificial way.

But no one animal model is capable of answering all the important questions related to the prevention or treatment of a particular type of human cancer. Researchers would be best served by turning their attention toward carefully crafting specific questions and letting the questions drive the selection of the model system. For some questions, cell culture or rodent studies will be appropriate. To answer others, researchers will have to resort to studying humans. In that sense, a human clinical trial is a form of animal model research—a specific collection of people is being used to represent the overall human population.   —D.J.W.

Advancing Cancer Therapy

VARIOUS CANCER TREATMENT studies featuring pet dogs have now been carried out or begun. Some of the earliest work focused on saving the limbs of teenagers with bone cancer. Twenty-five years ago a diagnosis of osteosarcoma in a youngster meant amputation of the affected limb, ineffective or no chemotherapy (drugs administered into the bloodstream to attack tumors anywhere in the body), and almost certain death. Today limb amputation can be avoided by chiseling out the diseased bone tissue and replacing it with a bone graft and metal implant—a process partially perfected in pet dogs by Stephen Withrow and his colleagues at Colorado State University. Withrow's team pioneered technical advances that reduced the likelihood of complications, such as placing bone cement in the marrow space of the bone graft. The researchers also showed that preoperative chemotherapy delivered directly into an artery could convert an inoperable tumor into an operable one. The group's work is credited with significantly increasing the percentage of teenagers who today can be cured of osteosarcoma.

Although a tumor's local effects are often controllable using surgery or radiation, metastasis is much harder to combat. For that, drug therapy is required. New compounds under development aim to disrupt key cellular events that regulate the survival and proliferation of metastatic tumor deposits as well as their sensitivity to cancer-fighting drugs. One experimental agent, ATN-161, which inhibits the formation of new blood vessels that foster tumor growth and metastasis, is currently being evaluated in large-breed dogs with bone cancers that have spread to the lungs. The ability of ATN-161 to enhance the effects of conventional chemotherapies is also under study. If these trials succeed, they could smooth the way toward clinical trials in humans.

Cancer researchers are also turning their attention to more familiar kinds of pharmaceuticals, including nonsteroidal anti-inflammatory drugs (NSAIDs), the class of compounds that includes ibuprofen. Certain NSAIDs have exhibited significant antitumor activity against a variety of canine tumors. In studies of pet dogs with bladder cancer, for example, the NSAID piroxicam showed such impressive antitumor activity that the drug is now in human clinical trials to see if this treatment can derail the progression of precancerous bladder lesions to life-threatening cancer.

Developing new cancer therapies is not just about finding novel drugs. It is about optimizing drug delivery to the patient. In your vein or up your nose? That is the kind of information scientists testing new agents against lung cancer need to know. If the right amount of drug does not make it to the tumor, then even substances with impressive credentials for killing tumor cells in a petri dish will not stand a chance of working in human patients. Moreover, delivering pharmaceuticals directly to the target—so-called regional therapy—has the added benefit of avoiding the toxicity associated with systemic therapy.

Investigators have used pet dogs to study the intranasal delivery of a cytokine, a small immune system molecule, called interleukin-2 (IL-2) to treat naturally occurring lung cancers. Positive results from these experiments led to feasibility trials of inhaled IL-2 in human patients with lung metastases, further leading to trials with another cytokine, granulocyte colony stimulating factor. Pet dogs can also aid researchers in optimizing the dosing and delivery protocols for drugs that have already made their way into human trials.

Another challenge that pet dogs are helping to overcome is determining the extent of tumor spread, called clinical staging. Accurate staging is critical for devising therapeutic game plans that will maximally benefit the patient while minimizing exposure to harsh treatments that are unlikely to help at a given disease stage. For example, the odds that a teenager will survive osteosarcoma are increased by accurate identification (and subsequent surgical removal) of lung metastases.

Doctors typically determine the presence and extent of such metastases with noninvasive imaging techniques, such as computed tomography (CT). To assess how accurate such scanning is, one of us (Waters), along with investigators from Indiana University School of Medicine, collected CT images of the lungs from pet dogs with metastatic bone cancer and then examined the tissue at autopsy to verify that what was interpreted as a "tumor" on the scan really was a tumor and not a mistake. Results showed that state-of-the-art imaging with CT—the same type used in clinical staging of bone cancer in teenagers—significantly underestimates the number of cancer deposits within the lung. By revealing the limited accuracy of existing and experimental techniques, pet dogs are helping optimize the next generation of technologies for improved cancer detection.

Taking Aim at Cancer Prevention

BUT CANCER RESEARCHERS are shooting for more than improved detection and better treatment; they also want to prevent the disease. Surprisingly, prevention is a relatively new concept within the cancer research community. What cardiologists have known for a long time—that millions of lives can be saved through the prevention of heart disease—is just now gaining traction in the cancer field. The term "chemoprevention" was coined 30 years ago to refer to the administration of compounds to prevent cancer, but scientists did not gather nationally to debate cutting-edge knowledge of cancer prevention until October 2002.

> Comparative oncologists ask, Why not transform the enormous amount of cancer in pet dogs into a national resource, both for helping other pets and aiding people?

Today the pace is quickening as investigators are examining a diverse armamentarium of potential cancer-protective agents. But finding the proper dose of promising agents has always been challenging. Indeed, failure to do so proved disastrous for some early human trials of preventives. For example, in two large lung cancer prevention trials, people receiving high doses of the antioxidant nutrient beta-carotene had an unexpected *increase* in lung cancer incidence compared with placebo-treated control subjects.

Can dogs accelerate progress in cancer prevention? Recently canine studies have helped define the dose of an antioxidant—the trace mineral selenium—that minimizes cancer-causing genetic damage within the aging prostate. The message from the dogs: when it comes to taking dietary supplements such as selenium to reduce your cancer risk, more of a good thing is not necessarily better. Elderly dogs given moderate doses ended up with less DNA damage in their prostates than dogs given lower or higher amounts. Comparative oncologists hold that dog studies conducted before large-scale human prevention trials are initiated can streamline the process of finding the most effective dose of cancer preventives and can enable oncologists to lob a well-aimed grenade at the cancer foe.

Pet dogs can assist in preventing human cancers in another way. For years, dogs in the research laboratory have advanced understanding of the acute and long-term effects of high doses of cancer-causing chemicals. But pet dogs, just by going about their daily lives, could serve as sentinels—watchdogs, if you will—to identify substances in our homes and in our backyards that are carcinogenic at lower doses. If something can cause cancer, the disease will show up in pets,

THE AUTHORS

DAVID J. WATERS and KATHLEEN WILDASIN share an interest in stimulating fresh thinking about cancer. Waters is professor of comparative oncology at Purdue University, associate director of the Purdue Center on Aging and the Life Course and executive director of the Gerald P. Murphy Cancer Foundation in West Lafayette, Ind. He earned his B.S. and D.V.M. at Cornell University and a Ph.D. in veterinary surgery at the University of Minnesota. Wildasin is a Kentucky-based medical and science writer.

with their compressed life spans, well before it will in people.

Take asbestos. Most human cases of mesothelioma (a malignancy of tissues lining the chest and abdomen) stem from asbestos exposure. Symptoms can appear up to 30 years after the incriminating exposure. Investigators have now documented that mesothelioma in pet dogs is also largely related to encountering asbestos, most likely through being near a master who came into contact with it through a hobby or work. But in dogs, the time between exposure and diagnosis is comparatively brief—less than eight years. So the appearance of the cancer in a dog can alert people to look for and remediate any remaining sources of asbestos. Also, closer monitoring of exposed individuals might lead to earlier diagnosis of mesothelioma and render these cancers curable.

Pet dogs could assist in discovering other environmental hazards. Some well-documented geographic "hot spots" show an unusually high incidence of certain cancers. For example, women living in Marin County, California, have the country's highest breast cancer rate. Scientists typically try to identify the factors contributing to cancer in hot spots by comparing the genetics and behavior of people who become afflicted and those who do not. To advance the effort, comparative oncologists are now establishing cancer registries for pet dogs in those areas. If both pets and people living in a particular community experience higher-than-normal cancer rates, the finding would strengthen suspicions that these malignancies are being triggered by something in the environment.

Analyzing tissues of dogs could even potentially speed identification of the specific hazard. Many toxic chemicals, such as pesticides, concentrate themselves in body fat. So it might make sense to collect tissues from dogs during common elective surgical procedures (for example, spaying) or at autopsy. Later, if an unusually high number of people in an area acquire a certain form of cancer, investigators could analyze levels of different chemicals in the samples to see if any are particularly prominent and worth exploring as a contributing factor.

Why Uncle Bill Avoided Cancer

BECAUSE CANCER IN PET DOGS is so commonplace, the animals might be able to assist in solving an age-old mystery.

Cancer Resistance: Lessons from the Oldest Old

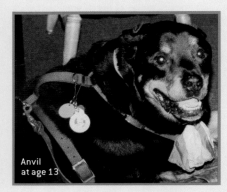

Anvil at age 13

Grace Fair at age 103

The risk of most human and canine cancers increases dramatically with age. This pattern has led to the belief that cancer is simply the result of a time-related accumulation of genetic damage. But recent studies of people who live to be 100 years old (centenarians) reveal an intriguing paradox: the oldest old among us are much less likely to succumb to cancer than are people who die in their 70s or 80s. Do the oldest-old pet dogs share a similar resistance to cancer mortality?

To answer this question, my colleagues and I consulted pet owners and veterinarians to construct lifetime medical histories of a large cohort of rottweiler dogs living in North America. We found that the likelihood of dying from cancer within two years rose with age during adulthood until dogs reached about 10 years but then declined after that. Moreover, exceptionally old dogs (those older than 13 years) were much less likely to die of cancer than were dogs with usual longevity even though the risk of dying from other causes continued to rise.

These findings raise the exciting possibility that studies comparing oldest-old dogs to those with usual longevity might reveal genes that regulate cancer resistance. Gene variations (so-called polymorphisms) responsible for cancer resistance and exceptional longevity in dogs could then be evaluated to see whether they are also overrepresented in the oldest-old humans. If they are, scientists can try to learn how the molecular interactions regulated by these genes alter cancer susceptibility at the tissue level.

At present, the precise nature of cancer resistance in human centenarians is poorly defined. Detailed autopsy studies of oldest-old dogs are currently under way to explore this issue. These studies should determine whether cancer resistance reflects a complete suppression of the biological events that give rise to cancer—for example, through increased repair of DNA damage—or whether tumors actually arise but are of the nonlife-threatening variety. By better understanding the genetic and pathological basis of cancer resistance in the oldest old, scientists will be better positioned to develop practical interventions that will reduce the average person's cancer risk. —D.J.W.

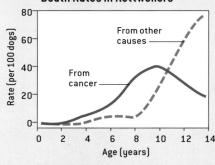

IN ROTTWEILERS who live past 10 years, getting older means having better odds of avoiding death from cancer.

build and reinforce alcohol addiction will most certainly help to better target existing treatments and devise new ones to break alcohol's hold.

Clues in Human Variations

GENES POWERFULLY INFLUENCE a person's physiology by giving rise to some 100,000 different types of protein, each of which has a direct role in the daily functioning of the body and brain or in regulating the activity of other genes. The strong connection between variations in basic physiology and an individual's susceptibility to alcohol problems is well illustrated by the very first gene to be identified as affecting the risk of developing alcohol dependence.

Decades ago researchers began investigating the widely observed tendency of persons from Chinese, Japanese or other East Asian backgrounds to become "flushed" when they drank an alcoholic beverage. Blood tests on subjects displaying this effect showed increased levels of acetaldehyde, a breakdown product of alcohol, which resulted in an uncomfortable sensation of warmth in the skin, palpitations and weakness. By the 1980s investigators traced the reaction to an enzyme involved in alcohol metabolism, aldehyde dehydrogenase, and eventually to the gene that encodes it, *ALDH1*. The enzyme breaks down acetaldehyde, but slight variations in the gene's DNA code in these subjects caused the enzyme to work more slowly. When these individuals ingested alcohol, the acetaldehyde—which may be toxic in high doses—was building up in their bodies.

This *ALDH1* gene variant has since been found to be common in Asian populations—seen in 44 percent of Japanese, 53 percent of Vietnamese, 27 percent of Koreans and 30 percent of Chinese (including 45 percent of Han Chinese)—yet it is rare in people of European descent. As might be expected, people with this slow-metabolizing gene variant also have a decreased risk, by up to sixfold, for alcoholism, so it is an example of a genetic variation that can protect against developing the disorder.

Other enzymes that break down alcohol have also been studied for their genetic contribution to alcohol dependence. Alcohol dehydrogenase (ADH), the enzyme responsible for the first step in the conversion of alcohol to acetaldehyde, for example, is actually produced by a family of genes, each of which affects different properties of the enzyme. The genes most important to alcohol metabolism are the *ADH1* group and *ADH4*. Our own studies of a U.S. population of European descent have recently provided strong evidence that variants in the *ADH4* genes in particular enhance the risk of alcoholism in members of that population, although exactly how these *ADH4* variants affect alcohol metabolism remains to be discovered.

Alcoholism is genetically complex, meaning that multiple genes are likely to be involved, and their interactions with one another and with an individual's environment also have to be examined before a complete picture of the processes that can lead to the disorder is assembled. People are also complex and manifest problems with alcohol in diverse ways, especially in the early stages of disease, although cases come to resemble one another clinically in the later stages of illness. Thus, when investigating the biology of alcoholism, researchers must carefully define the problem—for example, distinguishing between true dependence on alcohol and alcohol abuse, which is a less medically severe syndrome.

A widely used psychiatric standard for diagnosing dependence, be it on alcohol or another substance, requires that a person have experienced at least three of the following symptoms within the preceding 12 months: tolerance for large doses, withdrawal reactions, loss of control over use of the substance, efforts to stop or cut down, a great deal of time spent on the activity, giving up other activities, and continued use despite resulting physical or psychological problems. People who meet these criteria often have multiple cases of alcoholism in their families. With the willing participation of these subjects, we and other researchers have begun connecting individual symptoms with their physiological origins and ultimately with the responsible genes.

Indeed, an important strategy in the search for genes that affect a person's risk for alcohol dependence has been the examination of endophenotypes, which are physical traits—phenotypes—that are not externally visible but are measurable, and can therefore be studied to see whether certain patterns are more common in people with a complex disorder and may be linked to risk for that condition. The idea is grounded in an assumption that endophenotypes can reveal the biological bases for a disorder better

> People who meet criteria for dependence often have multiple cases of alcoholism in their families.

Overview/*Seeking Alcoholism Genes*

- Dependence on alcohol is a complex and controversial disorder, but susceptibility to it shows clear patterns of inheritance, which indicates that genes transmit some biological basis for greater vulnerability.
- Physiological traits, such as distinctive brain activity patterns in alcoholics and their children, help scientists pinpoint variant genes that affect a person's responses to alcohol.
- Finding the genes that influence alcoholism and related disorders provides insight into how the conditions develop, opens the way for better treatments, and allows individuals at high risk to make informed choices about their own health and behavior.

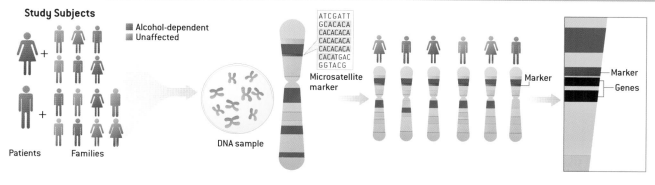

FINDING LINKS THROUGH FAMILIES

Identifying genes that influence a disorder as complex as alcoholism first involves linking the condition's traits to specific regions on chromosomes. This "linkage analysis" is easiest in genetically similar groups, such as families, with multiple members affected to some degree by the disorder.

Chromosomal features known as markers that appear more frequently in the affected relatives can flag potentially significant stretches of DNA. Detailed investigation of those regions can then reveal a gene whose function affects responses to alcohol.

RECRUITMENT
Alcoholics seeking treatment and their willing relatives are interviewed and diagnosed according to psychiatric criteria for alcohol dependence. All subjects provide DNA samples.

CHROMOSOME SURVEY
Researchers scan every person's chromosomes for patterns of repeated DNA known as microsatellite markers. In one individual, an alternating sequence of the bases cytosine and adenine might repeat 17 times, for example, whereas at the same location another relative has only 12 repeats of the sequence.

LINKAGE ANALYSIS
Markers frequently found in people with a specific trait of the disorder, but less often in unaffected kin, flag a chromosomal region linked to that trait.

GENE ASSOCIATION
Closer mapping of the DNA region near a marker reveals specific genes whose role in the disorder can be investigated.

than behavioral symptoms because they represent a fundamental physical trait that is more closely tied to its source in a gene variant. Although this approach to studying complex behaviors was first proposed in the 1970s by psychiatric researchers investigating schizophrenia, it has recently proved even more valuable with modern tools for assessing biologic processes and analyzing genetic data.

The brain's electrical activity patterns, for example, are a form of endophenotype. Using electroencephalography (EEG) to detect such activity through electrodes on the scalp, researchers can record patterns of neural firing. Sophisticated computer algorithms can analyze the data to identify the brain regions where the signals are likely to have originated, offering additional clues to the type of cognitive processing taking place. The overall brain waveforms and spikes in neural activity in response to specific stimuli seen in such EEG readings are distinctive in different individuals and serve as a kind of neurological fingerprint. These patterns can also reflect the general balance between excitatory processes within the brain, which render neurons more responsive to signaling from other neurons, and those that are inhibitory, making neurons less responsive.

Such electrophysiological patterns are highly heritable and they differ in characteristic ways in alcoholics and nonalcoholics, with excitation exceeding and overpowering inhibition in the alcoholic subjects' brains. This imbalance, or "disinhibition," can also be seen in the children of alcoholics and strongly predicts their own subsequent development of heavy drinking and alcohol dependence, which suggests that these patterns are a marker for a biologically inherited predisposition to alcoholism. Moreover, the signature patterns may point to the heritable vulnerability itself: disinhibition is believed to stem from a generalized lack of functioning inhibitory neurons in the brain areas responsible for judgment and decision making, and people lacking these inhibitory circuits may be more prone to acting on impulses originating in lower brain regions, such as the amygdala.

In the 1980s evidence from several laboratories showing that electrical activity in the brain could reveal a person's risk of alcohol dependence helped to stimulate the idea that an intensive search for the genes underpinning alcoholism-associated phenotypes was feasible and worthwhile. With support from the National Institute on Alcohol Abuse and Alcoholism, the Collaborative Study on the Genetics of Alcoholism (COGA), in which we are both participants, started in 1989. The study currently involves eight research centers across the U.S. and thousands of alcoholics and their family members who have agreed to help in this ongoing investigation.

Family Ties

AT COGA'S OUTSET, researchers at sites around the country sought to identify families severely affected by alcoholism. Previous twin, adoption and family studies had indicated that alcohol problems are strongly heritable—indeed, more than 50 percent of the overall risk for alcoholism is attributable to inherited factors, which makes family groups a powerful re-

SIGNATURES IN THE BRAIN

Certain patterns of brain electrical activity serve as measurable traits, known as endophenotypes, which reveal distinctive physiological characteristics of alcoholics and others at high risk for the disorder. Investigators have used these signature differences in brain function to uncover genes linked to alcoholism and related conditions.

The P300 Response
Measuring brain activity through electrodes on the scalp reveals a spike in signal strength (amplitude) between 300 and 500 milliseconds after a stimulus, such as a flash of light. Known as P300, this distinctive evoked response is significantly weaker in alcoholics, even when abstinent, than in nonalcoholics. A muted P300 is also typical in the children of alcoholic parents, indicating that this functional brain difference predates the onset of heavy drinking and is itself a risk factor for becoming alcoholic.

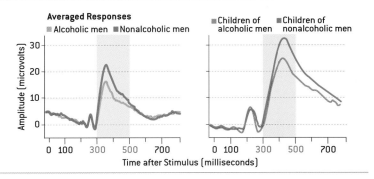

Dissecting the Response
P300 consists largely of neural signaling in low-frequency ranges known as delta and theta, which are associated with awareness and decision making. Mapping the EEG readings of nonalcoholic (*below*) and alcoholic (*bottom*) subjects by frequency reveals weaker signal strength in those bands among the alcoholics after 300 milliseconds. This trait was linked in family studies to both alcoholism and depression.

Linkage to a Gene
Reduced delta- and theta-frequency signal strength in alcoholic subjects was also traced to variants of *CHRM2*, a gene encoding a cellular receptor for the neurotransmitter acetylcholine, which regulates neural excitability.

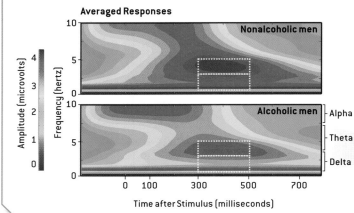
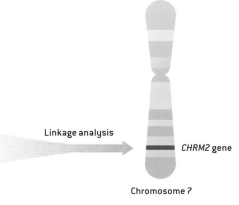

source for tracking specific traits and linking them to the relevant genes [*see box on preceding page*].

Some 1,200 subjects seeking treatment for alcohol dependence and their relatives—more than 11,000 people in all—were extensively interviewed. Among these, 262 families were found to be "deeply affected," which means that they included two or more first-degree relatives of the patient—such as parents or siblings—who were also diagnosed as alcohol-dependent. The electrophysiological brain endophenotypes of both affected and unaffected members of those families were assessed, and the subjects underwent further interviews to evaluate additional characteristics that are associated with alcoholism risk and believed to be genetically influenced. These traits include "low response," meaning that the person must consume larger-than-average amounts of alcohol before feeling its effects; previous experience of major depression; and certain drinking history patterns, such as a high maximum number of drinks ever consumed in a 24-hour period.

The participants also provided DNA samples, which allowed COGA scientists to examine the chromosomes of each individual and take note of distinctive molecular features, which can serve as markers for a potentially significant region of a chromosome. Markers appearing most frequently in family members exhibiting alcoholism-associated phenotypes would suggest a causal link between that region of a chromosome and the trait. Significant linkages were identified in this manner on chromosomes 1, 2, 4 and 7, and many years of genetic mapping subsequently pinpointed several specific genes in those regions, including *ADH4* and *GABRA2* on chromosome 4, as well as *CHRM2* on chromosome 7. Other research groups studying separate populations have also documented associations between a risk for alcoholism and these chromosomal regions and genes, confirming their likely role in the disorder.

Seeking the Connections: Alcoholism and our Genes
by John I. Nurnberger, Jr., and Laura Jean Bierut

BIOLOGY IN SOCIETY

1. If genetic profiling advances enough to allow the prediction of the risk of alcohol dependence, what measures should be taken if a 12-year-old boy, living with both a father and older sister who are alcoholics, is found to be genetically at high risk of alcohol dependence? Before answering, you may want to consider what additional information would be helpful in coming to a decision. Ultimately, your answer needs to consider the balance between parental legal rights, best prevention practices for the child, and respect for the family.

2. A variant of the hTAS2R16 bitter taste receptor gene is linked to alcoholism. This gene variant is present in almost 45% of African-Americans and is uncommon in Americans of European descent. How can this information be misused in a racist way in drawing conclusions about alcoholism in blacks and whites? What is the correct interpretation of this finding in understanding the risk of alcoholism in these two groups?

3. Alcohol abuse places a huge burden on the nation. For example, in 2005, 40% of traffic fatalities involved alcohol. If effective genetic screens for susceptibility to alcohol abuse become available, should they be mandatory for all citizens over the age of 15? Should people who test positive for alcohol susceptibility be legally barred from alcohol consumption?

THINKING ABOUT SCIENCE

1. Why are endophenotypes considered more helpful than behavioral phenotypes in searching for genes associated with alcoholism?

2. To what extent is an alcoholic responsible for his or her condition? Your answer should consider genetic and environmental factors in alcoholism and the idea that there are different paths to alcohol dependence.

3. The CHRM2 gene encodes a form of the acetylcholine receptor, and certain CHRM2 variants increase susceptibility to major depression and alcohol dependence. What are some possible relationships between major depression and alcoholism in individuals with the CHRM2 variant? Does the co-occurrence of major depression and alcoholism in some individuals with the variant CHRM2 gene prove that alcohol dependence and depression have the same underlying physiological cause?

WRITING ABOUT SCIENCE

Imagine that Dr. Nora Volkow, Director of the National Institute on Drug Abuse (NIDA), knows that you are an unbiased expert on alcoholism. She has been receiving conflicting advice about the most effective way of spending limited funds to reduce alcoholism in the United States. Some experts tell her that the value of future genetic screens for potential alcohol susceptibility genes, each of weak effect, is questionable, particularly when current programs to limit alcohol consumption and to treat alcoholics give immediate returns. Dr. Volkow asks you to prepare a brief (2- to 3-page) position paper. She asks that you include a discussion of the current state of the field, when you believe a genetic screen will be available, and the effectiveness of genetic screening for prevention and customized treatment of alcoholism. Finally, she wants you to recommend whether substantial funds from NIDA should be committed to research into the genetics of alcoholism.

Answers: Testing Your Comprehension
1 d; 2 d; 3 e; 4 c; 5 c; 6 d; 7 d; 8 c; 9 b; 10 a

RESTORING AMERICA'S

In the fall of 2004 a dozen conservation biologists gathered on a ranch in New Mexico to ponder a bold plan. The scientists, trained in a variety of disciplines, ranged from the grand old men of the field to those of us earlier in our careers. The idea we were mulling over was the reintroduction of large vertebrates—megafauna—to North America.

Most of these animals, such as mammoths and cheetahs, died out roughly 13,000 years ago, when humans from Eurasia began migrating to the continent. The theory—propounded 40 years ago by Paul Martin of the University of Arizona—is that overhunting by the new arrivals reduced the numbers of large vertebrates so severely that the populations could not recover. Called Pleistocene overkill, the concept was highly controversial at the time, but the general thesis that humans played a significant role is now widely accepted. Martin was present at the meeting in New Mexico, and his ideas on the loss of these animals, the ecological consequences, and what we should do about it formed the foundation of the proposal that emerged, which we dubbed Pleistocene rewilding.

Although the cheetahs, lions and mammoths that once roamed North America are extinct, the same species or close relatives have survived elsewhere, and our discussions focused on introducing these substitutes to North American ecosystems. We believe that these efforts hold the potential to partially restore important ecological processes, such as predation and browsing, to ecosystems where they have been absent for millennia. The substitutes would also bring economic and cultural benefits. Not surprisingly, the published proposal evoked strong reactions. Those reactions are welcome, because debate about the conservation issues that underlie Pleistocene rewilding merit thorough discussion.

Why Big Animals Are Important

OUR APPROACH concentrates on large animals because they exercise a disproportionate effect on the environment. For tens of millions of years, megafauna dominated the globe, strongly interacting and co-evolving with other species and influencing entire ecosystems. Horses, camels, lions, ele-

BIG, WILD ANIMALS

Pleistocene rewilding—a proposal to bring back animals that disappeared from North America 13,000 years ago—offers an optimistic agenda for 21st-century conservation

By C. Josh Donlan

phants and other large creatures were everywhere: megafauna were the norm. But starting roughly 50,000 years ago, the overwhelming majority went extinct. Today megafauna inhabit less than 10 percent of the globe.

Over the past decade, ecologist John Terborgh of Duke University has observed directly how critical large animals are to the health of ecosystems and how their loss adversely affects the natural world. When a hydroelectric dam flooded thousands of acres in Venezuela, Terborgh saw the water create dozens of islands—a fragmentation akin to the virtual islands created around the world as humans cut down trees, build shopping malls, and sprawl from urban centers. The islands in Venezuela were too small to support the creatures at the top of the food chain—predators such as jaguars, pumas and eagles. Their disappearance sparked a chain of reactions. Animals such as monkeys, leaf-cutter ants and other herbivores, whose populations were no longer kept in check by predation, thrived and subsequently destroyed vegetation—the ecosystems collapsed, with biodiversity being the ultimate loser.

Similar ecological disasters have occurred on other continents. Degraded ecosystems are not only bad for biodiversity; they are bad for human economies. In Central America, for instance, researchers have shown that intact tropical ecosystems are worth at least $60,000 a year to a single coffee farm because of the services they provide, such as the pollination of coffee crops.

Where large predators and herbivores still remain, they play pivotal roles. In Alaska, sea otters maintain kelp forest ecosystems by keeping herbivores that eat kelp, such as sea urchins, in check. In Africa, elephants are keystone players; as they move through an area, their knocking down trees and trampling create a habitat in which certain plants and animals can flourish. Lions and other predators control the populations of African herbivores, which in turn influence the distribution of plants and soil nutrients.

In Pleistocene America, large predators and herbivores played similar roles. Today most of that vital influence is absent. For example, the American cheetah (a relative of the African cheetah) dashed across the grasslands in pursuit of pronghorn antelopes for millions of years. These chases shaped the pronghorn's astounding speed and other biological aspects of one of the fastest animals alive. In the absence of the cheetah, the pronghorn appears "overbuilt" for its environment today.

Questions from Readers

A summary of this article on our Web site invited readers to ask questions about rewilding. Many of the questions helped to shape the article, and answers appear throughout the text. The author replies to a few others here.

Won't African elephants and lions freeze to death in the Montana winter? Will these animals be kept indoors in the winter, as they are in zoos? —Jason Raschka

The reader brings up an important point—that many questions, including climate suitability, would have to be answered by sound scientific studies during the process of Pleistocene rewilding. One could imagine scenarios in which animals would be free-living (albeit intensively managed) year-round in expansive reserves in the southwestern U.S. and perhaps other scenarios in locales farther north, where animals would be housed indoors during the coldest months. Even the latter alternative is arguably better than a zoo. Many zoos are struggling to find appropriate space for elephants and are choosing to abandon their elephant programs.

ELEPHANT at the Copenhagen zoo.

Aren't there big-game hunting ranches in Texas? Could we learn anything from them? —Foster

There are many big-game hunting ranches throughout Texas, and some of them hold animals such as cheetahs that we know, from the fossil record, once lived in North America. To my knowledge, no conservation biologists have studied the ranch-held animals, but these ranches might present excellent research opportunities if they are willing to collaborate. Other ranches, however, have animals that are not supported by the fossil record and as such offer no potential for study.

Hasn't this general idea been around for a while? —Kevin N.

Both the concept of rewilding and the term itself have been around for some time. Rewilding is the practice of reintroducing species to places from which they have been extirpated in the past few hundred years. Pleistocene rewilding, in contrast, involves introducing species descended from creatures that went extinct some 13,000 years ago or using similar species as proxies.

Overview/Rewilding Our Vision

- A group of conservationists has proposed reintroducing to North America large animals that went extinct 13,000 years ago.
- Close relatives of these animals—elephants, camels, lions, cheetahs—survived elsewhere; returning them to America would reestablish key ecological processes that once thrived there, as well as providing a refuge for endangered species from Africa and Asia and creating opportunities for ecotourism.
- The proposal has, understandably, evoked strong reactions, but the bold suggestion has spurred debate and put a positive spin on conservation biology, whose role has been mainly a struggle to slow the loss of biodiversity.

Pleistocene rewilding is not about recreating exactly some past state. Rather it is about restoring the kinds of species interactions that sustain thriving ecosystems. Giant tortoises, horses, camels, cheetahs, elephants and lions: they were all here, and they helped to shape North American ecosystems. Either the same species or closely related species are available for introduction as proxies, and many are already in captivity in the U.S. In essence, Pleistocene rewilding would help change the underlying premise of conservation biology from limiting extinction to actively restoring natural processes.

At first, our proposal may seem outrageous—lions in Montana? But the plan deserves serious debate for several reasons. First, nowhere on Earth is pristine, at least in terms of being substantially free of human influence. Our demographics, chemicals, economics and politics pervade every part of the planet. Even in our largest national parks, species go extinct without active intervention. And human encroachment shows alarming signs of worsening. Bold actions, rather than business as usual, will be needed to reverse such negative influences. Second, since conservation biology emerged as a discipline more than three decades ago, it has been mainly a business of doom and gloom, a struggle merely to slow the loss of biodiversity. But conservation need not be only reactive. A proactive approach would include restoring natural processes, starting with ones we know are disproportionately important, such as those influenced by megafauna.

Third, land in North America is available for the reintroduction of megafauna. Although the patterns of human land use are always shifting, in some areas, such as parts of the Great Plains and the Southwest, large private and public lands with low or declining human population densities might be used for the project. Fourth, bringing megafauna back to America would also bring tourist and other dollars into nearby communities and enhance the public's appreciation of the natural world. More than 1.5 million people visit San Diego's Wild Animal Park every year to catch a

When the West Was Really Wild

Soon after humans crossed the Bering land bridge into North America some 13,000 years ago, almost 75 percent of the continent's large mammals (those weighing more than 45 kilograms) disappeared (*color*). One of the goals of Pleistocene rewilding is to restore some of these species or close proxies to the American West. For example, the same species of lion and cheetah that once lived in North America survive today in Africa; the African or Asian elephant could substitute for the extinct mammoth; and Bactrian camels might stand in for the extinct *Camelops*.

Large Mammals of Pleistocene North America

Xenarthra
 Glyptodont (*Glyptotherium floridanum*)
 Harlan's ground sloth (*Paramylodon harlani*)
 Jefferson's ground sloth (*Megalonyx jeffersonii*)
 Shasta ground sloth (*Nothrotheriops shastensis*)

Carnivores (Carnivora)
 Dire wolf (*Canis dirus*)
 Gray wolf (*Canis lupus*)
 Black bear (*Ursus americanus*)
 Brown bear (*Ursus arctos*)
 Giant short-faced bear (*Arctodus simus*)
 Saber-toothed cat (*Smilodon fatalis*)
 American lion (*Panthera leo*)
 Jaguar (*Panthera onca*)
 American cheetah (*Miracinonyx trumani*)
 Mountain lion (*Puma concolor*)

Elephants (Proboscidea)
 American mastodon (*Mammut americanum*)
 Columbian mammoth (*Mammuthus columbi*)
 Dwarf mammoth (*Mammuthus exilis*)
 Woolly mammoth (*Mammuthus primigenius*)

Horses (Perissodactyla)
 Mexican horse (*Equus conversidens*)
 Western horse (*Equus occidentalis*)
 Other extinct horses and asses (*Equus spp.*)

Even-Toed Ungulates (Artiodactyla)
 Western camel (*Camelops hesternus*)
 Long-legged llama (*Hemiauchenia macrocephala*)
 Long-nosed peccary (*Mylohyus nasutus*)
 Flat-headed peccary (*Platygonus compressus*)
 Mule deer (*Odocoileus hemionus*)
 White-tailed deer (*Odocoileus virginianus*)
 Mountain deer (*Navahoceros fricki*)
 Woodland caribou (*Rangifer tarandus*)
 Moose (*Alces alces*)
 Wapiti (*Cervus elaphus*)
 Pronghorn (*Antilocapra americana*)
 Harrington's mountain goat (*Oreamnos harringtoni*)
 Mountain goat (*Oreamnos americanus*)
 Bighorn sheep (*Ovis canadensis*)
 Shrub ox (*Euceratherium collinum*)
 Bonnet-headed musk ox (*Bootherium bombifrons*)
 Bison (*Bison bison*)
 Extinct bison (*Bison spp.*)

Fadiga, found that answer somewhat accidentally in a surprising class of neurons in the monkey brain that fire when an individual performs simple goal-directed motor actions, such as grasping a piece of fruit. The surprising part was that these same neurons also fire when the individual sees someone else perform the same act. Because this newly discovered subset of cells seemed to directly reflect acts performed by another in the observer's brain, we named them mirror neurons.

Much as circuits of neurons are believed to store specific memories within the brain, sets of mirror neurons appear to encode templates for specific actions. This property may allow an individual not only to perform basic motor procedures without thinking about them but also to comprehend those acts when they are observed, without any need for explicit reasoning about them. John grasps Mary's action because even as it is happening before his eyes, it is also happening, in effect, inside his head. It is interesting to note that philosophers in the phenomenological tradition long ago posited that one had to experience something within oneself to truly comprehend it. But for neuroscientists, this finding of a physical basis for that idea in the mirror neuron system represents a dramatic change in the way we understand the way we understand.

Instant Recognition

OUR RESEARCH GROUP was not seeking to support or refute one philosophical position or another when we first noticed mirror neurons. We were studying the brain's motor cortex, particularly an area called F5 associated with hand and mouth movements, to learn how commands to perform certain actions are encoded by the firing patterns of neurons. For this purpose, we were recording the activity of individual neurons in the brains of macaques. Our laboratory contained a rich repertoire of stimuli for the monkeys, and as they performed various actions, such as grasping for a toy or a piece of food, we could see that distinct sets of neurons discharged during the execution of specific motor acts.

Then we began to notice something strange: when one of us grasped a piece of food, the monkeys' neurons would fire in the same way as when the monkeys themselves grasped the food. At first we wondered whether this phenomenon could be the result of some trivial factor, such as the monkey performing an unnoticed movement while observing our actions. Once we managed to rule out this possibility and others, including food expectation by the monkeys, we realized that the pattern of neuron activity associated with the observed action was a true representation in the brain of the act itself, regardless of who was performing it.

Often in biological research, the most direct way to establish the function of a gene, protein or group of cells is simply to eliminate it and then look for deficits in the organism's health or behavior afterward. We could not use this technique to determine the role of mirror neurons, however, because we found them spread across important regions on both sides of the brain, including the premotor and parietal cortices. Destroying the entire mirror neuron system would have produced such broad general cognitive deficits in the monkeys that teasing out specific effects of the missing cells would have been impossible.

So we adopted a different strategy. To test whether mirror neurons play a role in understanding an action rather than just visually registering it, we assessed the neurons' responses when the monkeys could comprehend the meaning of an action without actually seeing it. If mirror neurons truly mediate understanding, we reasoned, their activity should reflect the meaning of the action rather than its visual features. We therefore carried out two series of experiments.

First we tested whether the F5 mirror neurons could "recognize" actions merely from their sounds. We recorded the mirror neurons while a monkey was observing a hand motor act, such as ripping a sheet of paper or breaking a peanut shell, that is accompanied by a distinctive sound. Then we presented the monkey with the sound alone. We found that many F5 mirror neurons that had responded to the visual observation of acts accompanied by sounds also responded to the sounds alone, and we dubbed these cell subsets audiovisual mirror neurons.

Next we theorized that if mirror neurons are truly involved in understanding an action, they should also discharge when the monkey does not actually see the action but has sufficient clues to create a mental representation of it. Thus, we first showed a monkey an experimenter reaching for and grasping a piece of food. Next, a screen was positioned in front of the monkey so that it could not

> *The pattern of activity was a true representation in the brain of the act itself, regardless of who was performing it.*

Overview/Meeting of Minds

- Subsets of neurons in human and monkey brains respond when an individual performs certain actions and also when the subject observes others performing the same movements.
- These "mirror neurons" provide a direct internal experience, and therefore understanding, of another person's act, intention or emotion.
- Mirror neurons may also underlie the ability to imitate another's action, and thereby learn, making the mirror mechanism a bridge between individual brains for communication and connection on multiple levels.

REALITY REFLECTED

In experiments with monkeys, the authors discovered subsets of neurons in brain-motor areas (*right*) whose activation appeared to represent actions themselves. Firing by these "mirror neurons" could therefore produce in one individual an internal recognition of another's act. Because the neurons' response also reflected comprehension of the movement's goal, the authors concluded that action understanding is a primary purpose of the mirror mechanism. Involvement of the mirror neurons in comprehending the actor's final intention was also seen in their responses, which distinguished between identical grasping actions performed with different intentions.

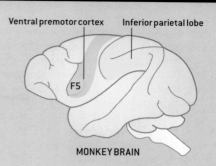

MONKEY BRAIN

UNDERSTANDING ACTION

In early tests, a neuron in the premotor area F5, associated with hand and mouth acts, became highly active when the monkey grasped a raisin on a plate (*1*). The same neuron also responded intensely when an experimenter grasped the raisin as the monkey watched (*2*).

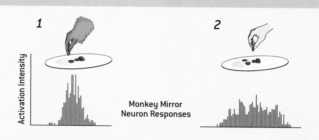

Monkey Mirror Neuron Responses

DISCRIMINATING GOAL

An F5 mirror neuron fired intensely when the monkey observed an experimenter's hand moving to grasp an object (*1*) but not when the hand motioned with no object as its goal (*2*). The same neuron did respond to goal-directed action when the monkey knew an object was behind an opaque screen, although the animal could not see the act's completion (*3*). The neuron responded weakly when the monkey knew no object was behind the screen (*4*).

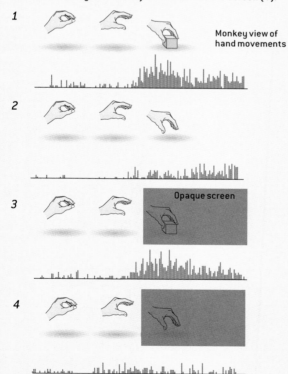

Monkey view of hand movements

DISCERNING INTENTION

In the inferior parietal lobe, readings from one neuron show intense firing when the monkey grasped a fruit to bring it to its mouth (*1*). The neuron's response was weaker when the monkey grasped the food to place it in a container (*2*). The same mirror neuron also responded intensely when the monkey watched an experimenter perform the grasp-to-eat gesture (*3*) and weakly to the grasp-to-place action (*4*). In all cases, the responses were associated with the grasping act, indicating that the neuron's initial activation encoded an understanding of final intention.

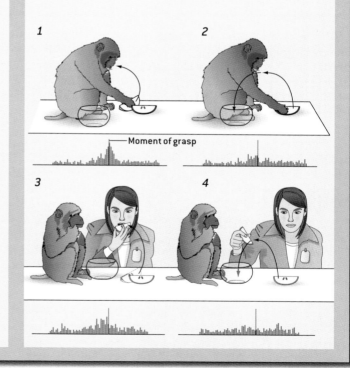

Moment of grasp

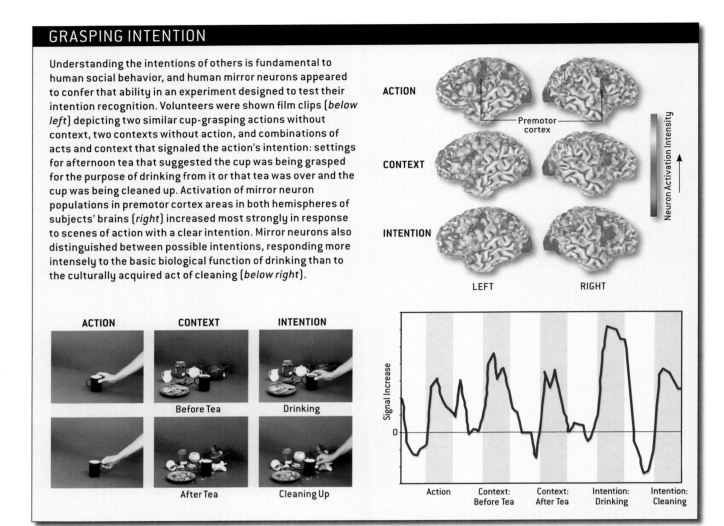

see the experimenter's hand grasping the food but could only guess the action's conclusion. Nevertheless, more than half the F5 mirror neurons also discharged when the monkey could just imagine what was happening behind the screen.

These experiments confirmed, therefore, that the activity of mirror neurons underpins understanding of motor acts: when comprehension of an action is possible on a nonvisual basis, such as sound or mental representation, mirror neurons do still discharge to signal the act's meaning.

Following these discoveries in the monkey brain, we naturally wondered whether a mirror neuron system also exists in humans. We first obtained strong evidence that it does through a series of experiments that employed various techniques for detecting changes in motor cortex activity. As volunteers observed an experimenter grasping objects or performing meaningless arm gestures, for example, increased neural activation in their hand and arm muscles that would be involved in the same movements suggested a mirror neuron response in the motor areas of their brains. Further investigations using different external measures of cortical activity, such as electroencephalography, also supported the existence of a mirror neuron system in humans. But none of the technologies we had used up to this point allowed us to identify the exact brain areas activated when the volunteers observed motor acts, so we set out to explore this question with direct brain-imaging techniques.

In those experiments, carried out at San Raffaele Hospital in Milan, we used positron-emission tomography (PET) to observe neuronal activity in the brains of human volunteers as they watched grasping actions performed with different hand grips and then, as a control, looked at stationary objects. In these situations, seeing actions performed by others activated three main areas of the brain's cortex. One of these, the superior temporal sulcus (STS), is known to contain neurons that respond to observations of moving body parts. The other two—the inferior parietal lobule (IPL) and the inferior frontal gyrus (IFG)—correspond, respectively, to the monkey IPL and the monkey ventral premotor cortex, including F5, the areas where we had previously recorded mirror neurons.

These encouraging results suggested a mirror mechanism at work in the human brain as well but still did not fully reveal its scope. If mirror neurons permit an observed act to be directly understood by experiencing it, for example, we wondered to what extent the ultimate goal of the action is also a component of that "understanding."

On Purpose

RETURNING TO our example of John and Mary, we said John knows both that Mary is picking up the flower and that she plans to hand it to him. Her smile gave him a contextual clue to her intention, and in this situation, John's knowledge of Mary's goal is fundamental to his understanding of her action, because giving him the flower is the completion of the movements that make up her act.

When we perform such a gesture ourselves, in reality we are performing a series of linked motor acts whose sequence is determined by our intent: one series of movements picks the flower and brings it to one's own nose to smell, but a partly different set of movements grasps the flower and hands it to someone else. Therefore, our research group set out to explore whether mirror neurons provide an understanding of intention by distinguishing between similar actions with different goals.

For this purpose, we returned to our monkeys to record their parietal neurons under varying conditions. In one set of experiments, a monkey's task was to grasp a piece of food and bring it to its mouth. Next we had the monkey grasp the same item and place it into a container. Interestingly, we found that most of the neurons we recorded discharged differently during the grasping part of the monkey's action, depending on its final goal. This evidence illustrated that the motor system is organized in neuronal chains, each of which encodes the specific intention of the act. We then asked whether this mechanism explains how we understand the intentions of others.

We tested the same grasping neurons for their mirror properties by having a monkey observe an experimenter performing the tasks the monkey itself had done earlier [see box on page 35]. In each instance, most of the mirror neurons were activated differently, depending on whether the experimenter brought the food to his mouth or put it in the container. The patterns of firing in the monkey's brain exactly matched those we observed when the monkey itself performed the acts—mirror neurons that discharged most strongly during grasping-to-eat rather than grasping-to-place did the same when the monkey watched the experimenter perform the corresponding action.

A strict link thus appears to exist between the motor organization of intentional actions and the capacity to understand the intentions of others. When the monkeys observed an action in a particular context, seeing just the first grasping component of the complete movement activated mirror neurons forming a motor chain that also encoded a specific intention. Which chain was activated during their observation of the beginning of an action depended on a variety of factors, such as the nature of the object acted on, the context and the memory of what the observed agent did before.

To see whether a similar mechanism for reading intentions exists in humans, we teamed with Marco Iacoboni and his colleagues at the University of California, Los Angeles, for a functional magnetic resonance imaging (fMRI) experiment on volunteers. Participants in these tests were presented with three kinds of stimuli, all contained within video clips. The first set of images showed a hand grasping a cup against an empty background using two different grips. The second consisted of two scenes containing objects such as plates and cutlery, arranged in one instance as though they were ready for someone to have afternoon tea and in the other as though they were left over from a previously eaten snack and were ready to be cleaned up. The third stimulus set showed a hand grasping a cup in either of those two contexts.

We wanted to establish whether human mirror neurons would distinguish between grasping a cup to drink, as suggested by the ready-for-tea context, and grabbing the cup to take it away, as suggested by the cleanup setting. Our results demonstrated not only that they do but also that the mirror neuron system responded strongly to the intention component of an act. Test subjects observing the hand motor acts in the "drinking" or "cleaning" contexts showed differing activation of their mirror neuron systems, and mirror neuron activity was stronger in both those situations than when subjects observed the hand grasping a cup without any context or when looking only at the place settings [see box on opposite page].

Given that humans and monkeys are social species, it is not difficult to see the potential survival advantage of a mechanism, based on mirror neurons, that locks basic motor acts onto a larger motor semantic network, permitting the direct and immediate comprehension of others' behavior without complex cognitive machinery. In social life, however, understanding others' emotions is equal-

> *When people use the expression "I feel your pain," they may not realize how literally it could be true.*

THE AUTHORS

GIACOMO RIZZOLATTI, LEONARDO FOGASSI and VITTORIO GALLESE work together at the University of Parma in Italy, where Rizzolatti is director of the neurosciences department and Fogassi and Gallese are associate professors. In the early 1990s their studies of motor systems in the brains of monkeys and humans first revealed the existence of neurons with mirror properties. They have since continued to investigate those mirror neurons in both species as well as the role of the motor system in general cognition. They frequently collaborate with the many other research groups in Europe and the U.S. now also studying the breadth and functions of the mirror neuron system in humans and animals.

ly important. Indeed, emotion is often a key contextual element that signals the intent of an action. That is why we and other research groups have also been exploring whether the mirror system allows us to understand what others feel in addition to what they do.

Connect and Learn

AS WITH ACTIONS, humans undoubtedly understand emotions in more than one way. Observing another person experiencing emotion can trigger a cognitive elaboration of that sensory information, which ultimately results in a logical conclusion about what the other is feeling. It may also, however, result in direct mapping of that sensory information onto the motor structures that would produce the experience of that emotion in the observer. These two means of recognizing emotions are profoundly different: with the first, the observer deduces the emotion but does not feel it; via the second, recognition is firsthand because the mirror mechanism elicits the same emotional state in the observer. Thus, when people use the expression "I feel your pain" to indicate both comprehension and empathy, they may not realize just how literally true their statement could be.

A paradigmatic example is the emotion of disgust, a basic reaction whose expression has important survival value for fellow members of a species. In its most primitive form, disgust indicates that something the individual tastes or smells is bad and, most likely, dangerous. Once again using fMRI studies, we collaborated with French neuroscientists to show that experiencing disgust as a result of inhaling foul odorants and witnessing disgust on the face of someone else activate the same neural structure—the anterior insula—at some of the very same locations within that structure [*see box below*]. These results indicate that populations of mirror neurons in the insula become active both when the test participants experience the emotion and when they see it expressed by others. In other words, the observer and the observed share a neural mechanism that enables a form of direct experiential understanding.

Tania Singer and her colleagues at University College London found similar matches between experienced and observed emotions in the context of pain. In that experiment, the participants felt pain produced by electrodes placed on their hands and then watched electrodes placed on a test partner's hand followed by a cue for painful stimulation. Both situations activated the same regions of the anterior insula and the anterior cingulate cortex in the subjects.

Taken together, such data strongly suggest that humans may comprehend emotions, or at least powerful negative emotions, through a direct mapping mechanism involving parts of the brain that generate visceral motor responses. Such a mirror mechanism for understanding emotions cannot, of course, fully explain all social cognition, but it does provide for the first time a functional neural basis for some of the interpersonal relations on which more complex social behaviors are built. It may be a substrate that allows us to empathize with others, for example. Dysfunction in this mirroring system may also be implicated in empathy deficits, such as those seen in children with autism [see "Broken Mirrors: A Theory of Autism," by Vilayanur S. Ramachandran and Lindsay M. Oberman, SCIENTIFIC AMERICAN, November 2006].

Many laboratories, including our own, are continuing to explore these questions, both for their inherent interest and their potential therapeutic applications. If the mirror neuron template of a motor action is partly inscribed in the brain by experience, for instance, then it should theoretically be possible to alleviate motor impairments, such as those suffered following a stroke, by potentiating undamaged action templates. Recent evidence indicates, in fact, that the mirror mechanism also plays a role in the way we initially learn new skills.

Although the word "ape" is often used to denote mimicry, imitation is not an especially well developed ability among nonhuman primates. It is rare in monkeys and limited in the great apes, including chimpanzees and gorillas. For

EMOTIONAL MIRRORS

Feeling disgust activated similar parts of the brain when human volunteers experienced the emotion while smelling a disgusting odor or when the same subjects watched a film clip (*left*) of someone else disgusted. In this brain cross section, neuron populations activated by the experience of disgust are outlined in red, and those activated by seeing disgust are circled in yellow. (Blue outlines the region of investigation, and green indicates areas examined in a previous study.) These overlapping neuron groups may represent a physical neural mechanism for human empathy that permits understanding the emotions of others.

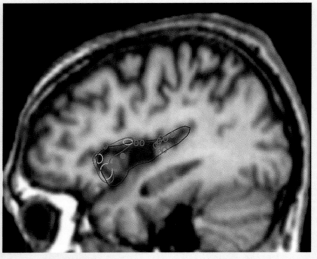

IMITATION requires reproduction of actions performed by another person. If mirror neurons underlie the uniquely human facility for imitation, the mirror system may serve as a bridge that allows us to teach and learn new skills.

humans, in contrast, imitation is a very important means by which we learn and transmit skills, language and culture. Did this advance over our primate relatives evolve on the neural substrate of the mirror neuron system? Iacoboni and his group provided the first evidence that this might be the case when they used fMRI to observe human subjects who were watching and imitating finger movements. Both activities triggered the IFG, part of the mirror neuron system, in particular when the movement had a specific goal.

In all these experiments, however, the movements to be imitated were simple and highly practiced. What role might mirror neurons play when we have to learn completely new and complex motor acts by imitation? To answer this question, Giovanni Buccino at our university and collaborators in Germany recently used fMRI to study participants imitating guitar chords after seeing them played by an expert guitarist. While test subjects observed the expert, their parietofrontal mirror neuron systems became active. And the same area was even more strongly activated during the subjects' imitation of the chord movements. Interestingly, in the interval following observation, while the participants were programming their own imitation of the guitar chords, an additional brain region became active.

Known as prefrontal area 46, this part of the brain is traditionally associated with motor planning and working memory and may therefore play a central role in properly assembling the elementary motor acts that constitute the action the subject is about to imitate.

Many aspects of imitation have long perplexed neuroscientists, including the basic question of how an individual's brain takes in visual information and translates it to be reproduced in motor terms. If the mirror neuron system serves as a bridge in this process, then in addition to providing an understanding of other people's actions, intentions and emotions, it may have evolved to become an important component in the human capacity for observation-based learning of sophisticated cognitive skills.

Scientists do not yet know if the mirror neuron system is unique to primates or if other animals possess it as well. Our own research group is currently testing rats to see if that species also demonstrates mirror neuron responses. Such internal mirroring may be an ability that developed late in evolution, which would explain why it is more extensive in humans than in monkeys. Because even newborn human and monkey babies can imitate simple gestures such as sticking out the tongue, however, the ability to create mirror templates for observed actions could be innate. And because lack of emotional mirroring ability appears to be a hallmark of autism, we are also working with young autistic children to learn whether they have detectable motor deficits that could signal a general dysfunction of the mirror neuron system.

Only a decade has passed since we published our first discoveries about mirror neurons, and many questions remain to be answered, including the mirror system's possible role in language—one of humanity's most sophisticated cognitive skills. The human mirror neuron system does include Broca's area, a fundamental language-related cortical center. And if, as some linguists believe, human communication first began with facial and hand gestures, then mirror neurons would have played an important role in language evolution. In fact, the mirror mechanism solves two fundamental communication problems: parity and direct comprehension. Parity requires that meaning within the message is the same for the sender as for the recipient. Direct comprehension means that no previous agreement between individuals—on arbitrary symbols, for instance—is needed for them to understand each other. The accord is inherent in the neural organization of both people. Internal mirrors may thus be what allow John and Mary to connect wordlessly and permit human beings in general to communicate on multiple levels. 🅂🄰

MORE TO EXPLORE

Action Recognition in the Premotor Cortex. Vittorio Gallese, Luciano Fadiga, Leonardo Fogassi and Giacomo Rizzolatti in *Brain*, Vol. 119, No. 2, pages 593–609; April 1996.

A Unifying View of the Basis of Social Cognition. V. Gallese, C. Keysers and G. Rizzolatti in *Trends in Cognitive Sciences*, Vol. 8, pages 396–403; 2004.

Grasping the Intentions of Others with One's Own Mirror Neuron System. Marco Iacoboni et al. in *PLoS Biology*, Vol. 3, Issue 3, pages 529–535; March 2005.

Parietal Lobe: From Action Organization to Intention Understanding. Leonardo Fogassi et al. in *Science*, Vol. 302, pages 662–667; April 29, 2005.

Questions for Review

TEST YOUR COMPREHENSION

1. Mirror neurons are involved in understanding an action and
   a. performing that action.
   b. storing memories.
   c. allowing reasoning.
   d. inhibiting socially unacceptable behaviors.
   e. controlling breathing.

2. A role of mirror neurons in understanding actions was established by
   a. recording neural activity across the entire brain.
   b. training monkeys to perform a task like grasping a container of food.
   c. seeing whether specific mirror neurons fire when the action is performed.
   d. seeing whether specific mirror neurons fire when the action is observed.
   e. seeing whether specific mirror neurons fire when the action is imagined.

3. Why couldn't the authors deduce mirror neuron function from careful observation of the effects of eliminating mirror neurons?
   a. No information on the location of mirror neurons was available.
   b. Mirror neurons cannot be eliminated without destroying brain function.
   c. There are no methods available for eliminating neurons from the brain.
   d. Others had already published results using this approach.
   e. The authors did not wish to use an untested technique of inferring function by eliminating cells.

4. The finding that monkey mirror neurons fire when animals can imagine but not actually see an experimenter grasping food provided evidence that
   a. mirror neurons are essential for motor activity.
   b. mirror neurons do not have a direct role in motor activity.
   c. mirror neurons are activated only by basic biological functions.
   d. mirror neurons are involved in understanding actions.
   e. mirror neurons occur in many different areas of the brain.

5. Experiments in which human volunteers viewed a cup being grasped in the context of a place setting for tea or a set of dishes after a meal tested the idea that
   a. mirror neurons fire when an action is observed.
   b. the same set of mirror neurons fire in both the actor and the observer.
   c. mirror neurons are involved in understanding the intention of acts.
   d. mirror neurons are particularly active in the inferior parietal lobule and inferior frontal gyrus.
   e. mirror neurons in the same regions of human and monkey brains are stimulated by food-related actions.

6. Mirror neurons add physiological meaning to the phrase
   a. you can't have your cake and eat it too.
   b. seize the day.
   c. if I knew then what I know now.
   d. you can't win them all.
   e. I feel your pain.

7. Mirror neurons are involved in understanding
   a. thought processes.
   b. emotions.
   c. health.
   d. social status.
   e. intelligence.

8. If particular circuits of mirror neurons are established by experience, then this gives hope for helping victims of
   a. stroke.
   b. amyotrophic lateral sclerosis (Lou Gehrig's disease).
   c. cancer.
   d. Down syndrome.
   e. multiple sclerosis.

9. The mirror neuron function that is most highly evolved in humans is
   a. understanding intent of motor actions.
   b. understanding emotion.
   c. rational thinking.
   d. imitation.
   e. spirituality.

10. A disorder that may involve a deficit in mirror neurons is
    a. autism.
    b. Down syndrome.
    c. schizophrenia.
    d. paranoia.
    e. depression.

Mirrors in the Mind by Giacomo Rizzolatti, Leonardo Fogassi, and Vittorio Gallese

BIOLOGY IN SOCIETY

1. Initial knowledge of mirror neurons came from experiments in which electrodes were implanted into the brains of living monkeys. The insights from these experiments are profound and have the potential to explain what goes wrong in autism and to help stroke victims. Could this knowledge have been gained without experiments using living animals? Given what's been learned and the potential applications of this knowledge, are these experimental procedures justified?

2. Three new video game systems were released in 2006. Two of these use traditional game controllers, and the other requires the player to use body motions that mimic those of game characters or those used in sports such as tennis, golf, and snowboarding. Could the manufacturer of the motion-demanding controller fairly market the product as a controller that more effectively stimulates mirror neurons?

3. Imagine that you've been hired at a leading advertising agency. How might understanding mirror neurons help you to design effective ads? Would knowledge of mirror neuron function be more helpful in marketing some products, say potato chips, than in others, perhaps life insurance? Further imagine that, although your boss thought you were crazy when you told her about mirror neurons, she says, "You're on. Give me an ad for life insurance that uses this mirror neuron angle." What does your TV ad for life insurance look like? How does it use the "mirror neuron angle"?

THINKING ABOUT SCIENCE

1. The authors discuss two ways in which an individual might understand the intent of someone performing an action. What are these two ways of understanding? If there was no experimental evidence favoring one method over the other, which way would you argue for, considering the immediacy of understanding of intent?

2. Scientists were amazed when they first noticed that the same set of motor neurons fired in the same patterns both in brains of monkeys performing an action and in those observing that action. The mirror neuron hypothesis immediately came to mind. What else might explain parallel firing patterns in the doer and observer? How would you test these alternative hypotheses?

3. The authors hypothesize that if mirror neurons are involved in understanding an action, then they should become active when there are cues to create a mental representation of the action, even if it is not directly observed. Design an experiment that tests whether mirror neurons are involved in a monkey's understanding of intent when a seated investigator working with another monkey stands up, turns, and begins walking away. Assume that mirror neurons have been shown to be activated when the monkey observes this action. Your experimental design should focus on how to test mirror neuron involvement in understanding by the monkey when the action is not seen.

WRITING ABOUT SCIENCE

How near are we to having a physical explanation for consciousness? In what ways are questions of modern neuroscience and traditional philosophy converging? If the understanding of another person's intent can be explained by the firing patterns of mirror neurons, does this denigrate human understanding? Consider these or related questions as you write an essay about how modern neuroscience is uncovering the way the mind works and how this understanding fits with long-standing philosophical views of human consciousness. You should support your arguments at least in part with findings described in the article.

Answers: Testing Your Comprehension
1a; 2e; 3b; 4d; 5c; 6e; 7b; 8a; 9d; 10a

NUTRITION

EATING MADE SIMPLE

How do you cope with a mountain of conflicting diet advice?

By Marion Nestle

KEY CONCEPTS

- Nutrition advice is confusing. Scientists have difficulty deriving clear guidelines because a study of an individual nutrient fails to produce an understanding of what happens to it when mixed with other nutrients in the body.

- The picture becomes more clouded because industry groups constantly press their message to both government agencies and consumers about the benefits of eating particular food products.

- The simplest message may be the best: do not overeat, exercise more, consume mostly fruits, vegetables and whole grains, and avoid junk foods.

—*The Editors*

As a nutrition professor, I am constantly asked why nutrition advice seems to change so much and why experts so often disagree. Whose information, people ask, can we trust? I'm tempted to say, "Mine, of course," but I understand the problem. Yes, nutrition advice seems endlessly mired in scientific argument, the self-interest of food companies and compromises by government regulators. Nevertheless, basic dietary principles are not in dispute: eat less; move more; eat fruits, vegetables and whole grains; and avoid too much junk food.

"Eat less" means consume fewer calories, which translates into eating smaller portions and steering clear of frequent between-meal snacks. "Move more" refers to the need to balance calorie intake with physical activity. Eating fruits, vegetables and whole grains provides nutrients unavailable from other foods. Avoiding junk food means to shun "foods of minimal nutritional value"—highly processed sweets and snacks laden with salt, sugars and artificial additives. Soft drinks are the prototypical junk food; they contain sweeteners but few or no nutrients.

If you follow these precepts, other aspects of the diet matter much less. Ironically, this advice has not changed in years. The noted cardiologist Ancel Keys (who died in 2004 at the age of

OVERABUNDANCE of food choices confronts shoppers and diners every day.

ORGANIC FOODS have been shown to leave people who eat them with fewer synthetic pesticides in their bodies than are found in those who consume conventional foods. Proving that organics contain more vitamins or antioxidants is more difficult, but preliminary studies clearly suggest that they do.

OLD ADVICE STILL HOLDS TRUE

In 1959 Ancel and Margaret Keys offered the following—familiar and still useful—precepts regarding nutrition and activity:

- Do not get fat; if you are fat, reduce.
- Restrict saturated fats: fats in beef, pork, lamb, sausages, margarine and solid shortenings; fats in dairy products.
- Prefer vegetable oils to solid fats but keep total fats under 30 percent of your diet calories.
- Favor fresh vegetables, fruits and nonfat milk products.
- Avoid heavy use of salt and refined sugar.
- Good diets do not depend on drugs and fancy preparations.
- Get plenty of exercise and outdoor recreation.

100) and his wife, Margaret, suggested similar principles for preventing coronary heart disease nearly 50 years ago [see sidebar at left].

But I can see why dietary advice seems like a moving target. Nutrition research is so difficult to conduct that it seldom produces unambiguous results. Ambiguity requires interpretation. And interpretation is influenced by the individual's point of view, which can become thoroughly entangled with the science.

Nutrition Science Challenges

This scientific uncertainty is not overly surprising given that humans eat so many different foods. For any individual, the health effects of diets are modulated by genetics but also by education and income levels, job satisfaction, physical fitness, and the use of cigarettes or alcohol. To simplify this situation, researchers typically examine the effects of single dietary components one by one.

Studies focusing on one nutrient in isolation have worked splendidly to explain symptoms caused by deficiencies of vitamins or minerals. But this approach is less useful for chronic conditions such as coronary heart disease and diabetes that are caused by the interaction of dietary, genetic, behavioral and social factors. If nutrition science seems puzzling, it is because researchers typically examine single nutrients detached from food itself, foods separate from diets, and risk factors apart from other behaviors. This kind of research is "reductive" in that it attributes health effects to the consumption of one nutrient or food when it is the overall dietary pattern that really counts most.

For chronic diseases, single nutrients usually alter risk by amounts too small to measure except through large, costly population studies. As seen recently in the Women's Health Initiative, a clinical trial that examined the effects of low-fat diets on heart disease and cancer, participants were unable to stick with the restrictive dietary protocols. Because humans cannot be caged and fed measured formulas, the diets of experimental and control study groups tend to converge, making differences indistinguishable over the long run—even with fancy statistics.

It's the Calories

Food companies prefer studies of single nutrients because they can use the results to sell products. Add vitamins to candies, and you can market them as health foods. Health claims on the labels of junk foods distract consumers from their caloric content. This practice matters because when it comes to obesity—which dominates nutrition problems even in some of the poorest countries of the world—it is the calories that count. Obesity arises when people consume significantly more calories than they expend in physical activity.

America's obesity rates began to rise sharply in the early 1980s. Sociologists often attribute the "calories in" side of this trend to the demands of an overworked population for convenience foods—prepared, packaged products and restaurant meals that usually contain more calories than home-cooked meals.

But other social forces also promoted the calorie imbalance. The arrival of the Reagan administration in 1980 increased the pace of industry deregulation, removing controls on agricultural production and encouraging farmers to grow more food. Calories available per capita in the national food supply (that produced by American farmers, plus imports, less exports) rose from 3,200 a day in 1980 to 3,900 a day two decades later [see box on opposite page].

The early 1980s also marked the advent of the "shareholder value movement" on Wall Street. Stockholder demands for higher short-term returns on investments forced food companies to expand sales in a marketplace that already contained excessive calories. Food companies responded by seeking new sales and marketing opportunities. They encouraged formerly shunned practices that eventually changed social norms, such as frequent between-meal snacking, eating in book and clothing stores, and serving larger portions. The industry continued to sponsor organizations and journals that focus on nutrition-related subjects and intensified its efforts to lobby government for favorable dietary advice. Then and now food lobbies have promoted positive interpretations of scientific studies, sponsored research that can be used as a basis for health claims, and attacked critics, myself among them, as proponents of "junk science." If anything, such activities only add to public confusion.

FOOD FACTOIDS

To reduce your weight by a pound of fat a week, eat 500 fewer calories each day.

Carbohydrates and proteins have about 4 calories per gram. Food fats contain more than twice as much: 9 calories per gram. A teaspoon holds about 5 grams.

Alcohol is metabolized in a way that promotes accumulation of fat in the liver, leading to the proverbial beer belly.

An adult expends about 100 calories for every mile walked or run. It takes nearly three miles to burn off the calories in a 20-ounce soft drink.

Supermarkets as "Ground Zero"

No matter whom I speak to, I hear pleas for help in dealing with supermarkets, considered by shoppers as "ground zero" for distinguishing health claims from scientific advice. So I spent a year visiting supermarkets to help people think more clearly about food choices. The result was my book *What to Eat*.

Supermarkets provide a vital public service but are not social services agencies. Their job is to sell as much food as possible. Every aspect of store design—from shelf position to background music—is based on marketing research [*see center item on page 50*]. Because this research shows that the more products customers see, the more they buy, a store's objective is to expose shoppers to the maximum number of products they will tolerate viewing.

[OBESITY GAINS]
AS FOOD CALORIES SWELL, SO DO WAISTLINES

A substantial rise in U.S. obesity rates during the past few decades was paralleled by increases in the availability of larger portion sizes, total calories, caloric sweeteners and sugary soft drinks in the food supply. The apparent dip in three of these measures (calories, sugars and sugary soft drinks) after 1998 may be explained by greater use of artificial sweeteners and the partial replacement of sugary soft drinks with beverages that are not sweetened with sugars.

U.S. OBESITY RATES ON THE RISE
Percent of total population (ages 20–74) classified as obese

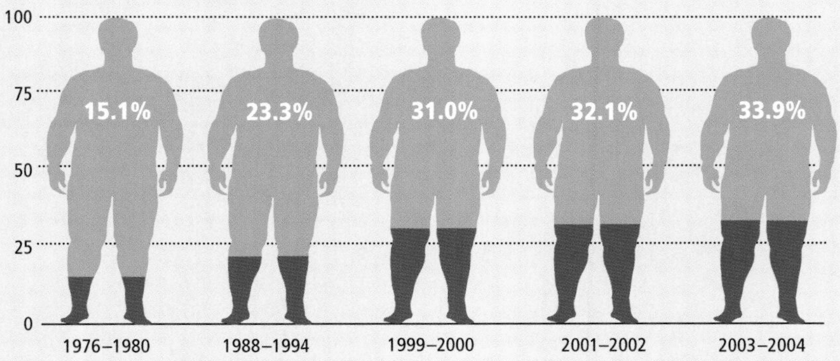

- 1976–1980: 15.1%
- 1988–1994: 23.3%
- 1999–2000: 31.0%
- 2001–2002: 32.1%
- 2003–2004: 33.9%

SUPER-SIZE PORTIONS GROW
Number of food items introduced in larger sizes by restaurants and manufacturers in the U.S.

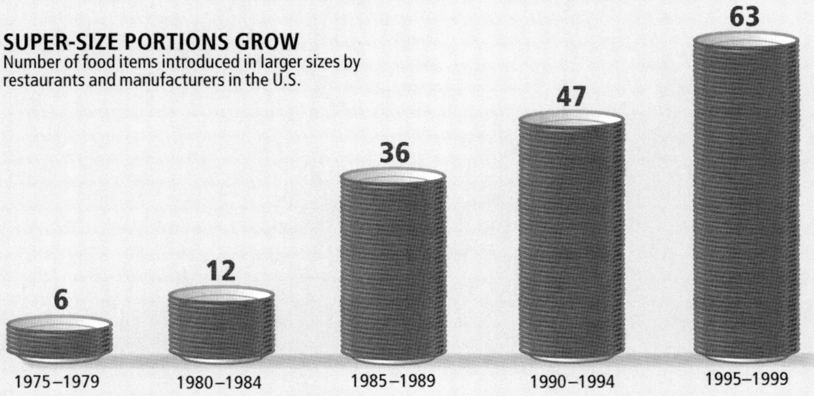

- 1975–1979: 6
- 1980–1984: 12
- 1985–1989: 36
- 1990–1994: 47
- 1995–1999: 63

CALORIES AVAILABLE
Per person per day in the U.S. food supply

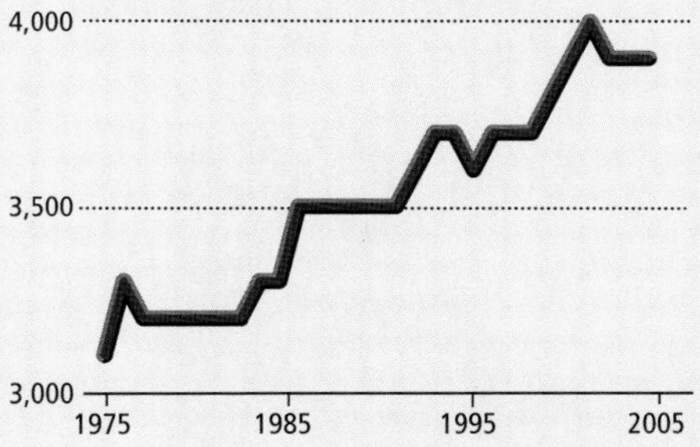

CALORIC SWEETENERS AVAILABLE
Pounds per person in the U.S. food supply

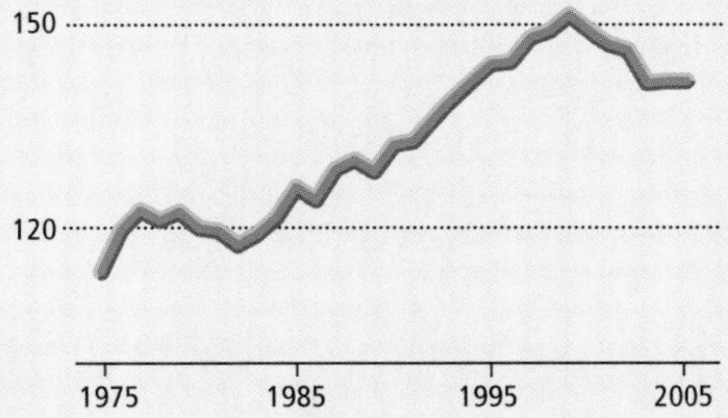

SUGARY SOFT DRINKS AVAILABLE
Gallons per person in the U.S. food supply

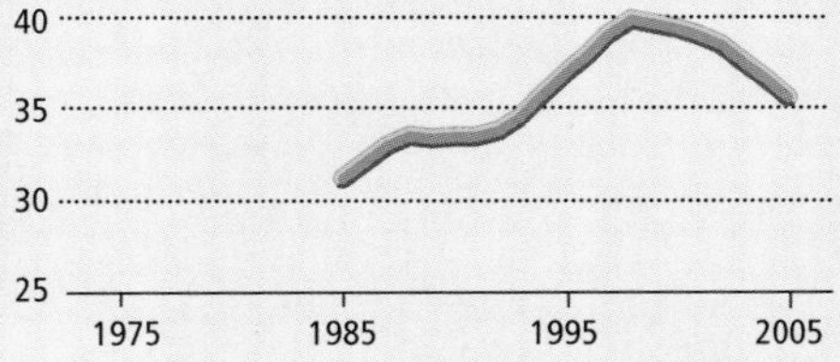

If consumers are confused about which foods to buy, it is surely because the choices require knowledge of issues that are not easily resolved by science and are strongly swayed by social and economic considerations. Such decisions play out every day in every store aisle.

Are Organics Healthier?

Organic foods are the fastest-growing segment of the industry, in part because people are willing to pay more for foods that they believe are healthier and more nutritious. The U.S. Department of Agriculture forbids producers of "Certified Organic" fruits and vegetables from using synthetic pesticides, herbicides, fertilizers, genetically modified seeds, irradiation or fertilizer derived from sewage sludge. It licenses inspectors to ensure that producers follow those rules. Although the USDA is responsible for organics, its principal mandate is to promote conventional agriculture, which explains why the department asserts that it "makes no claims that organically produced food is safer or more nutritious than conventionally produced food. Organic food differs from conventionally grown food in the way it is grown, handled and processed."

This statement implies that such differences are unimportant. Critics of organic foods would agree; they question the reliability of organic certification and the productivity, safety and health benefits of organic production methods.

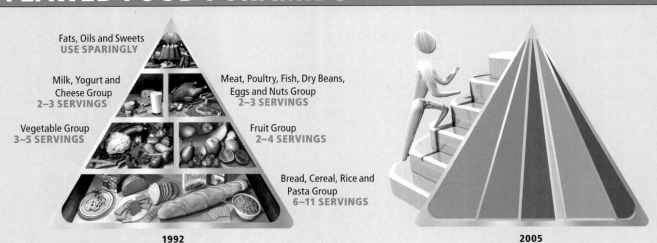

[GOVERNMENT ADVICE]
FLAWED FOOD PYRAMIDS

1992 / **2005**

Whether you found the food pyramid created by the U.S. Department of Agriculture in 1992 beneficial or not, it was at least simple to use. The familiar triangular nutrition guide suggested how much of each food category—grains, dairy products, fruits and vegetables, meats and fats, oils and sweets—one should eat every day.

But in my opinion, the USDA's 2005 replacement, MyPyramid, is a disaster. The process the agriculture agency employed to replace the 1992 food pyramid (*left*) has been kept secret. It remains a mystery, for example, just how the department came up with a design for a new food guide that emphasizes physical activity but is devoid of food (*right*). According to the USDA staff, people should keep physically active, eat in moderation, make personalized food choices, eat a variety of foods in the recommended number of servings, and pursue gradual dietary improvement. The color and width of the vertical bands of MyPyramid are meant to denote food groups and servings, but the only way to know this in detail is to log on to a computer. Users must go to www.pyramid.gov and type in gender, age and activity level to obtain a "personalized" dietary plan at one of a dozen calorie levels.

People who seek advice from this site, and millions have, find diet plans notable for the large amounts of food they seem to recommend and for the virtual absence of appeals to "eat less" or to "avoid" certain foods. Critics, not surprisingly, discern the strong influence of food industry lobbyists here. I myself, for example, am expected to consume four cups of fruits and vegetables, six ounces of grains, five ounces of meat and, of course, three cups of milk a day, along with a couple of hundred "discretionary calories" that I can spend on junk foods. For all its flaws, the 1992 pyramid was easier to understand and use.

What MyPyramid really lacks is any notion of a hierarchical ranking of the items in a single food group in terms of nutritional desirability. The preliminary design of MyPyramid in 2004 looked much like the final version with one critical exception: it illustrated a hierarchy of desirable food choices. The grain band, for instance, placed whole-grain bread at the bottom (a positive ranking), pasta about halfway up (a middle rank) and cinnamon buns at the top ("eat less"). In the final version, the USDA eliminated all traces of hierarchy, presumably because food companies do not want federal agencies to advise eating less of their products, useful as such recommendations might be to an overweight public. —*M.N.*

SOME OBESE CHILDREN in the U.S. consume 1,000 calories a day (equal to about 59 sugar cubes) in sweetened beverages alone. That is equivalent to more than three 20-ounce bottles of soft drinks.

Meanwhile the organic food industry longs for research to address such criticisms, but studies are expensive and difficult to conduct. Nevertheless, existing research in this area has established that organic farms are nearly as productive as conventional farms, use less energy and leave soils in better condition. People who eat foods grown without synthetic pesticides ought to have fewer such chemicals in their bodies, and they do. Because the organic rules require pretreatment of manure and other steps to reduce the amount of pathogens in soil treatments, organic foods should be just as safe—or safer—than conventional foods.

Similarly, organic foods ought to be at least as nutritious as conventional foods. And proving organics to be more nutritious could help justify their higher prices. For minerals, this task is not difficult. The mineral content of plants depends on the amounts present in the soil in which they are grown. Organic foods are cultivated in richer soils, so their mineral content is higher.

But differences are harder to demonstrate for vitamins or antioxidants (plant substances that reduce tissue damage induced by free radicals); higher levels of these nutrients relate more to a food plant's genetic strain or protection from unfavorable conditions after harvesting than to production methods. Still, preliminary studies show benefits: organic peaches and pears contain greater quantities of vitamins C and E, and organic berries and corn contain more antioxidants.

Further research will likely confirm that organic foods contain higher nutrient levels, but it is unclear whether these nutrients would make a measurable improvement in health. All fruits and vegetables contain useful nutrients, albeit in different combinations and concentrations. Eating a variety of food plants is surely more important to health than small differences in the nutrient content of any one food. Organics may be somewhat healthier to eat, but they are far less likely to damage the environment, and that is reason enough to choose them at the supermarket.

Dairy and Calcium

Scientists cannot easily resolve questions about the health effects of dairy foods. Milk has many components, and the health of people who consume milk or dairy foods is influenced by everything else they eat and do. But this area of research is especially controversial because it affects an industry that vigorously promotes dairy products as beneficial and opposes suggestions to the contrary.

Dairy foods contribute about 70 percent of the calcium in American diets. This necessary mineral is a principal constituent of bones, which constantly lose and regain calcium during normal metabolism. Diets must contain enough calcium to replace losses, or else bones become prone to fracture. Experts advise consumption of at least one gram of calcium a day to replace everyday losses. Only dairy foods provide this much calcium without supplementation.

But bones are not just made of calcium; they require the full complement of essential nutrients to maintain strength. Bones are stronger in people who are physically active and who do not smoke cigarettes or drink much alcohol. Studies examining the effects of single nutrients in dairy foods show that some nutritional factors—magnesium, potassium, vitamin D and lactose, for example—promote calcium re-

[THE AUTHOR]

Marion Nestle is Paulette Goddard Professor in the department of nutrition, food studies and public health and professor of sociology at New York University. She received a Ph.D. in molecular biology and an M.P.H. in public health nutrition from the University of California, Berkeley. Nestle's research focuses on scientific and social factors that influence food choices and recommendations. She is author of *Food Politics* (2002, revised 2007), *Safe Food* (2003) and *What to Eat* (2006).

tention in bones. Others, such as protein, phosphorus and sodium, foster calcium excretion. So bone strength depends more on overall patterns of diet and behavior than simply on calcium intake.

Populations that do not typically consume dairy products appear to exhibit lower rates of bone fracture despite consuming far less calcium than recommended [*see sidebar on opposite page*]. Why this is so is unclear. Perhaps their diets contain less protein from meat and dairy foods, less sodium from processed foods and less phosphorus from soft drinks, so they retain calcium more effectively. The fact that calcium balance depends on multiple factors could explain why rates of osteoporosis (bone density loss) are highest in countries where people eat the most dairy foods. Further research may clarify such counterintuitive observations.

In the meantime, dairy foods are fine to eat if you like them, but they are not a nutritional requirement. Think of cows: they do not drink milk after weaning, but their bones support bodies weighing 800 pounds or more. Cows feed on grass, and grass contains calcium in small amounts—but those amounts add up. If you eat plenty of fruits, vegetables and whole grains, you can have healthy bones without having to consume dairy foods.

A Meaty Debate

Critics point to meat as the culprit responsible for elevating blood cholesterol, along with raising risks for heart disease, cancer and other

DROPPING WEIGHT... AND KEEPING IT OFF

By Paul Raeburn

This past March, Stanford University researchers published the results of one of the longest and most persuasive comparisons of weight-loss programs ever conducted. Three of the four diets in the study are heavily promoted regimens that have made their originators famous: the Atkins diet and the Zone diet, which both emphasize high-protein foods, and the Ornish diet, a plan that prohibits most fatty foods. The fourth was the no-frills, low-fat diet that most nutrition experts recommend.

The results, published in the *Journal of the American Medical Association,* were a surprise because they seemed to overturn the conventional wisdom. The experts' low-fat diet was beaten by Atkins's steak dinners and bacon-and-egg breakfasts. A year after starting their diets, people on the Atkins plan—which unapologetically endorses high-fat protein such as meats and dairy products to keep dieters sated—had dropped an average of 10 pounds. Subjects on the other diets had lost between three and six pounds (*graph on opposite page*). And members of the Atkins test group showed no jump in blood cholesterol levels, despite the high levels of cholesterol in their diet.

Reporters jumped on the obvious headlines: "Atkins Fares Best..." stated the *Washington Post.* "Atkins Beats Zone, Ornish and U.S. Diet Advice," the Associated Press declared. It was the same everywhere else: Atkins had bested the competition.

The newspaper accounts were not wrong. But the lead author of the Stanford study suggests a different interpretation of the findings. "What happened in our study was very modest weight loss in all four groups," says Christopher D. Gardner, a nutrition scientist at the Stanford Prevention Research Center. All groups also showed improvement in individuals' levels of cholesterol, blood pressure and insulin, even though none of them followed their diet plans exactly. And far from overturning established ideas about low-fat diets, the Stanford investigation provided resounding confirmation of another generally held belief: most people who try to lose weight, on any kind of diet, will succeed, even if many of them regain the weight later.

Contrast those conclusions with the results of another study published in the April issue of *American Psychologist* by researchers at the University of California, Los Angeles. They analyzed 31 long-term diet studies and found, as Gardner said, that most participants did see results—losing about 5 to 10 percent of their total body mass. And they did it while on all kinds of diets. But most also regained all that weight over the longer term, and some put on even more than they had lost. Only a small minority of subjects in the 31 studies kept the extra pounds off. The researchers' conclusion? Eat in moderation and exercise regularly. (This statement parallels similar advice nutritionist Marion Nestle presents in the accompanying article.)

Gardner thinks the traditional exhortation to cut dietary fat has turned out to be a bad message. The public health experts got it wrong, he says: "It totally backfired on us." People who consumed less fat often turned to soda and similar corn-syrup-sweetened products, along with other refined, low-fiber, carbohydrate-rich foods. As a result, "the obesity epidemic has continued to grow. Calories have continued to creep up, and it's been predominantly in the refined carbohydrates."

YO-YO DIETING is unhealthy.

conditions. Supporters cite the lack of compelling science to justify such allegations; they emphasize the nutritional benefits of meat protein, vitamins and minerals. Indeed, studies in developing countries demonstrate health improvements when growing children are fed even small amounts of meat.

But because bacteria in a cow's rumen attach hydrogen atoms to unsaturated fatty acids, beef fat is highly saturated—the kind of fat that increases the risk of coronary heart disease. All fats and oils contain some saturated fatty acids, but animal fats, especially those from beef, have more saturated fatty acids than vegetable fats. Nutritionists recommend eating no more than a heaping tablespoon (20 grams) of saturated fatty acids a day. Beef eaters easily meet or exceed this limit. The smallest McDonald's cheeseburger contains 6 grams of saturated fatty acids, but a Hardee's Monster Thickburger has 45 grams.

Why meat might boost cancer risks, however, is a matter of speculation. Scientists began to link meat to cancer in the 1970s, but even after decades of subsequent research they remain unsure if the relevant factor might be fat, saturated fat, protein, carcinogens or something else related to meat. By the late 1990s experts could conclude only that eating beef probably increases the risk of colon and rectal cancers and possibly enhances the odds of acquiring breast, prostate and perhaps other cancers. Faced with this uncertainty, the American Cancer Society suggests selecting leaner cuts, smaller portions

The Atkins plan, which advises dieters to be less concerned about fat, steers people toward vegetables and protein and away from sugars and refined carbohydrates. "Maybe low carb is a better simple message to the public than low fat," Gardner says. "We tell them low carb, and they get it. They cut out a couple of sodas or a couple of cookies, and that adds up."

James Hill, a psychologist and authority on weight loss, agrees that the Atkins approach has virtues. "The Atkins diet is a great way to lose weight," he says. But it "is not a way to keep weight off," he asserts. "There's no way you can do it forever."

Hill is not terribly interested in comparing diets or devising new ones. "I think the weight-loss part is something we do pretty well," he says. One of his areas of research concerns individuals who have reduced their weight and sustained it. Hill and Rena Wing of Brown University have established what they call the National Weight Control Registry to collect data on people who have cut at least 30 pounds and kept them off for a year. Many have lost much more—the average is a 70-pound weight loss maintained for six years. "If you look at how they lost weight, there's no commonality at all," Hill says. But "if you look at how they kept it off, there's a lot of commonality."

The key, he continues, is exercise. "Activity becomes the driver; food restriction doesn't do it. The idea that for the rest of your life you're going to be hungry all the time—that's just silly." People in the registry get an average of an hour of physical activity every day, with some exercising for as much as 90 minutes a day. They also keep the fat in their diet relatively low, at about 25 percent of their calorie intake. Nearly all of them eat breakfast every day, and they weigh themselves regularly. "They tell us two things," Hill says. "The quality of life is higher—life is better than it was before." And "they get to the point with physical activity where they don't say they love it, but they say 'it's part of my life.'"

Hill admits that fitting an hour or more of exercise into the day is difficult, which is why he also focuses on prevention. Many of these people might never have become obese initially if they had exercised a mere 15 to 20 minutes a day. "I think you pay a price for having been obese," he states, "and you have to do a lot of activity to make up for that." ■

Paul Raeburn writes about science, policy and the environment from New York City.

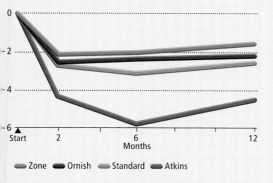

BATTLE OF THE DIET PLANS
Mean weight change over time (in kilograms)

● Zone ● Ornish ● Standard ● Atkins

CONTRARY TO EXPECTATIONS, the high-fat Atkins diet produced greater losses than three other popular weight-reduction plans.

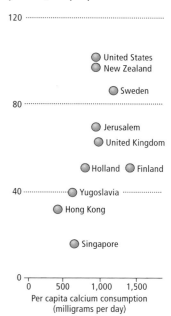

NOT MILK?

Surprisingly, some populations that eat few calcium-rich milk products appear in some descriptive studies to have lower rates of hip fractures than others that consume large quantities of dairy foods, despite the fact that diets of the former group contain far less calcium than experts recommend. This observation has not been fully explained.

Calcium intake and hip fractures

Incidence of hip fractures per 100,000 people

OMEGA-3 FATTY ACIDS, which are thought by some researchers to protect against heart disease, are found in oily fish such as salmon and trout.

and alternatives such as chicken, fish or beans—steps consistent with today's basic advice about what to eat.

Fish and Heart Disease

Fatty fish are the most important sources of long-chain omega-3 fatty acids. In the early 1970s Danish investigators observed surprisingly low frequencies of heart disease among indigenous populations in Greenland that typically ate fatty fish, seals and whales. The researchers attributed the protective effect to the foods' content of omega-3 fatty acids. Some subsequent studies—but by no means all—confirm this idea.

Because large, fatty fish are likely to have accumulated methylmercury and other toxins through predation, however, eating them raises questions about the balance between benefits and risks. Understandably, the fish industry is eager to prove that the health benefits of omega-3s outweigh any risks from eating fish.

Even independent studies on omega-3 fats can be interpreted differently. In 2004 the National Oceanic and Atmospheric Administration—for fish, the agency equivalent to the USDA—asked the Institute of Medicine (IOM) to review studies of the benefits and risks of consuming seafood. The ensuing review of the research on heart disease risk illustrates the challenge such work poses for interpretation.

The IOM's October 2006 report concluded that eating seafood reduces the risk of heart disease but judged the studies too inconsistent to decide if omega-3 fats were responsible. In contrast, investigators from the Harvard School of Public Health published a much more positive report in the *Journal of the American Medical Association* that same month. Even modest consumption of fish omega-3s, they stated, would cut coronary deaths by 36 percent and total mortality by 17 percent, meaning that not eating fish would constitute a health risk.

Differences in interpretation explain how distinguished scientists could arrive at such different conclusions after considering the same studies. The two groups, for example, had conflicting views of earlier work published in March 2006 in the *British Medical Journal*. That study found no overall effect of omega-3s on heart disease risk or mortality, although a subset of the original studies displayed a 14 percent reduction in total mortality that did not reach statistical significance. The IOM team interpreted the "nonsignificant" result as evidence for the need for caution, whereas the Harvard group saw the data as consistent with studies reporting the benefits of omega-3s. When studies present inconsistent results, both interpretations are plausible. I favor caution in such situations, but not everyone agrees.

Because findings are inconsistent, so is dietary advice about eating fish. The American Heart Association recommends that adults eat fatty fish at least twice a week, but U.S. dietary guidelines say: "Limited evidence suggests an association between consumption of fatty acids in fish and reduced risks of mortality from cardiovascular disease for the general population ... however, more research is needed." Whether or not fish uniquely protects against heart disease, seafood is a delicious source of many nutrients, and two small servings per week of the less predatory classes of fish are unlikely to cause harm.

Sodas and Obesity

Sugars and corn sweeteners account for a large fraction of the calories in many supermarket foods, and virtually all the calories in drinks—soft, sports and juice—come from added sugars.

In a trend that correlates closely with rising rates of obesity, daily per capita consumption of sweetened beverages has grown by about 200 calories since the early 1980s. Although common sense suggests that this increase might have something to do with weight gain, beverage makers argue that studies cannot prove that sugary drinks alone—independent of calories or other foods in the diet—boost the risk of obesi-

DESIGNER SUPERMARKETS

Marketing experts design nearly every feature of food stores—from product placement to mood music—to maximize sales.

When customers enter a grocery store, the first thing they see is typically something colorful, aromatic and enticing—fresh produce, for example.

The long center aisles and aisle-end displays are jam-packed with products, forcing shoppers to pass by many items that they might purchase on impulse.

Food companies pay supermarkets to get their products—salty chips and other junk foods—positioned prominently in huge displays.

Checkout lines are plastered with candy and other junk food items—the last temptation.

CLIMATE CHANGE

The surprising recent finding that living plants produce methane does not throw doubt on the cause of global warming. Human activities—not plants—are the source of the surge in this and other greenhouse gases

By Frank Keppler and Thomas Röckmann

What do you do as a scientist when you discover something that clearly contradicts the textbooks? The two of us faced this problem head-on when experiments we were running in 2005 showed that living vegetation produces the greenhouse gas methane. The established view held that only microbes that thrive without oxygen (anaerobic bacteria) can manufacture this gas. But our tests unexpectedly revealed that green plants also make methane—and quite a lot of it.

The first thing we did was look for errors in our experimental design and for every conceivable scenario that could have led us astray. Once we satisfied ourselves that our results were valid, though, we realized we had come across something very special, and we began to think about the consequences of our findings and how to present them to other researchers. Difficult as this discovery had been for us to accept, trying to convince our scientific peers and the public was almost impossible—in large part because we had to explain how such an important source of methane could have been overlooked for decades by the many able investigators studying methane and puzzling over climate change.

Natural Gas

MOST PEOPLE KNOW methane (often written as the chemical formula CH_4) as natural gas. Found in oil fields and coal beds as well as in natural gas fields, it has become an important source of energy and will most likely remain so given the limited reserves of oil on the planet. Approximately 600 million metric tons of it—both anthropogenic (from human activities) and natural—rise into the atmosphere every year. Most of these emissions have been thought to come from the decay of nonfossil organic material as a result of activity by anaerobic bacteria. Wetlands such as swamps, marshes and rice paddies provide the greatest share. Cattle, sheep and termites also make methane, as a by-product of anaerobic microbial digestion in their gut. Forest and savanna fires release methane, as does the combustion of fossil fuels [see box on page 57]. Over the years, researchers have gained considerable knowledge about the global methane cycle, and the consensus of the Intergovernmental Panel on Climate Change (IPCC) in 2001 was that the major sources had probably been identified (although the proportion each source contributes was still uncertain).

Nevertheless, some observations were difficult to explain. For instance, large fluctuations of atmospheric methane during the ice ages and warm ages, which have been reconstructed from air bubbles trapped in ice cores, remained a mystery. But no scientist in 2001 would have factored in direct emissions of methane by plants, because no one suspected that biological production of methane by anything other than microbial anaerobic processes was possible.

Knowing the sources of methane and how much they emit is important because methane is an extremely efficient greenhouse gas. Much more carbon dioxide is spewed into the atmosphere every year, but one kilogram of methane warms the earth 23 times more than a kilogram of carbon dioxide does. As a result of human activities, the concentration of methane in the atmosphere has almost tripled over the past 150 years. Will it continue to increase into the 21st century? Can emissions be reduced? Climate scientists need to answer such questions, and to do so we must know the origin and fate of this important gas.

Startling Findings

THE IDEA OF INVESTIGATING plants as methane emitters grew out of research we had been conducting on chloromethane, a chlorinated gas that destroys ozone and was thought to come mainly from the oceans and forest fires. A few years ago, while working at the Department of Agriculture and Food Science in Northern Ireland, we discovered that aging plants provide most of the chloromethane found in the atmosphere. Because methane, like chloromethane, is released during the burning of biomass, we wondered whether intact plants might also release methane.

To satisfy our curiosity, we collected 30 different kinds of tree leaves and grasses from tropical and temperate regions and placed them in small chambers with typical concentrations of atmospheric oxygen. To our amazement, all of the various kinds of leaves and plant litter produced methane. Usually a gram of dried plant material releases between 0.2 and three nanograms (one billionth of a gram) of methane an hour. These relatively tiny amounts were difficult to monitor, even using our highly sensitive state-of-the-art equipment.

The task was made still more challenging because we had to differentiate between methane produced by plant tissue and the high background levels normally present in ambient air. We believe this difficulty is what prevented biologists from observing the phenomenon earlier. The secret to our discovery was that we removed the interfering effect of the natural methane background by flushing the chambers with methane-free air before the start of each experiment. We were then able to measure the methane released by plant tissue.

Our curiosity fueled, we undertook similar experiments with living plants [*see box on page 58*], and we found that the rates of methane production increased dramatically, jumping to 10 to 100 times those of leaves detached from plants. By running a series of experiments, we excluded the possibility that bacteria that thrive without oxygen produced the methane. Finally, we were absolutely convinced that living plants release methane in significant quantities. We could provide no immediate answers about the mechanism of how they did this, although we suspect that pectin, a substance in the walls of the plant cells, is involved. We decided that this question would have to await further research, which is currently under way. Because of methane's role in climate change, however, we realized it was crucial to begin to take into account the quantity of gas released into the atmosphere by this newly discovered source.

How much might plants be contributing to the planet's methane totals? It was immediately obvious to us that even though a single leaf or plant made only tiny amounts of methane, these small bits would add up quickly because plants cover a substantial part of the globe. We were nonetheless astounded by the figure generated by our calculations: between 60 million and 240 million metric tons of methane come from plants every year—this constitutes 10 to 40 percent of annual global emissions. Most of it, about two thirds, originates in the vegetation-rich tropics. We knew, of course, that extrapolating global estimates from a limited sample of laboratory measurements was open to error. Still, the final number seemed extremely large—and if it surprised us, it would be heresy to many of our scientific peers.

Fortunately for us, support for our work soon came from an unexpected source. A group of environmental physicists in Heidelberg, Germany, was observing the earth's atmosphere from space. In 2005 the scientists' satellite measurements revealed "clouds" of methane over tropical forests [*see illustration on page 59*]. They reported that their observa-

> We collected 30 different kinds of tree leaves and grasses. To our AMAZEMENT all of the various kinds of leaves and plant litter produced METHANE.

Overview/*Nature's Surprise*

- The established view has been that methane (natural gas) is produced by microbes that thrive without oxygen, but experiments by the authors' team unexpectedly revealed that living plants also manufacture this potent greenhouse gas.
- Although this startling finding can explain many previously puzzling observations, a number of scientists are still skeptical, in particular about the amount of methane that plants generate. Knowing the sources of methane and how much they emit is important because of methane's role in trapping heat.
- An early misinterpretation of the finding suggested that forests might actually be contributing to global warming, but the authors emphasize that plants do not contribute to the recent increase in methane and global warming.

THE TEXTBOOK VIEW

In the past 150 years, methane emissions into the atmosphere have roughly tripled (*graph*), and today some 600 million metric tons are sent into the air annually. That rise is a concern because methane, like carbon dioxide, traps heat in the earth's atmosphere and therefore contributes to global warming.

Until the authors and their colleagues published their recent discoveries, traditional thinking held that all natural releases of methane resulted from the activity of bacteria that thrive in wet, oxygen-poor environments. Such environments include swamps and rice paddies as well as the digestive systems of termites and ruminants. And analyses of the sources of the gas in the environment (*pie charts*) indicated that the dramatic rise in methane concentrations since the mid-1800s has stemmed from human industrial activities (such as the use of fossil fuels for energy) and increased rice cultivation and breeding of ruminants (because of population growth). The authors' work casts no doubt on the explanation for why methane concentrations in the atmosphere have increased, but estimates of the relative contributions to methane levels from natural sources will have to be revised.

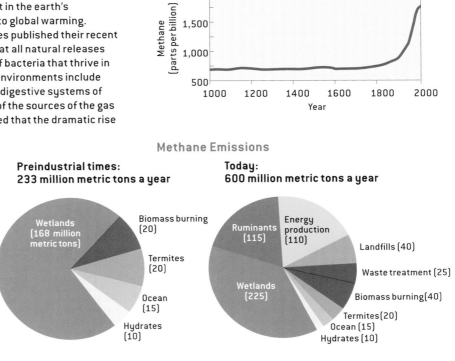

tions could not be explained by simply using the current understanding of the global methane budget. In light of our findings, however, their work made sense: green vegetation was the source of the methane clouds.

Recently further support has come from Paul J. Crutzen, a 1995 Nobel Prize winner, and his colleagues. After our findings were published in January 2006, they reanalyzed measurements made in 1988 of air samples from the Venezuelan savanna and concluded that 30 million to 60 million metric tons of methane could be released from vegetation in these regions. Crutzen said that "looking back to 1988, we could have made the discovery, but accepting the general wisdom that methane can only be produced under anaerobic conditions, we missed the boat."

Despite this support for our work, many scientists are still skeptical about methane emissions from plants, especially about our estimate of how much methane comes from vegetation. A number of our scientific colleagues are therefore recalculating the budget for the plant source, using different methods from ours but applying our emission rates. Of course, we keenly await an independent verification of our laboratory findings.

Solving an Old Puzzle

OUR FINDINGS WOULD EXPLAIN a trend that has puzzled climate scientists for years: fluctuations in methane levels in parallel with changes in global temperatures. Ice cores serve as natural archives that store information about atmospheric composition and climate variability going back almost a million years. Tiny bubbles of air trapped in the ice reveal the relative concentrations of atmospheric gases in the past [*see box on next page*]. We see in the ice cores, for example, that variations of past carbon dioxide levels are closely linked to changes in global temperatures. During ice ages, carbon dioxide concentrations are low; during warm spells, levels increase.

In general, methane concentrations follow the same trend as carbon dioxide, but the reason has been unclear. Scientists have tried to use models of wetlands (the only major natural source of methane previously believed to exist) to reconstruct the curious variations of past methane levels. Yet they found it difficult to reproduce the reported differences in atmospheric methane levels between glacial and interglacial periods.

THE AUTHORS

FRANK KEPPLER and THOMAS RÖCKMANN first discovered methane emissions from plants when they were working together at the Max Planck Institute for Nuclear Physics in Heidelberg, Germany. Keppler earned a Ph.D. in environmental geochemistry from the University of Heidelberg in 2000. He recently received a European Young Investigator Award (EURYI) to build his own research group at the Max Planck Institute for Chemistry in Mainz. Röckmann received his Ph.D. from the University of Heidelberg. In 2005 he was appointed full professor at the Institute for Marine and Atmospheric Research Utrecht in the Netherlands.

THE NEW VIEW

The authors' team scrutinized the gases emitted by plant debris and by living plants. To their surprise, the scientists found that both plant debris and growing vegetation produce methane. This important source of emissions had been overlooked until the team performed experiments in chambers that had been flushed of methane, which allowed the researchers to measure the minute amounts of the gas that plants give off.

The new view could explain puzzling fluctuations in methane levels that mirror changes in levels of carbon dioxide and in global temperatures (*graphs*). Scientists have tracked these changes by studying ice cores, in which trapped bubbles preserve information about the composition of the atmosphere going back almost a million years; concentrations of deuterium in the ice provide information about temperature. High atmospheric carbon dioxide concentrations and rising temperatures most likely led to a large increase in vegetation, which could have been accompanied by correspondingly large releases of methane.

AUTHORS' EXPERIMENT detected minuscule quantities of methane produced by living vegetation (*rye grass in photograph*).

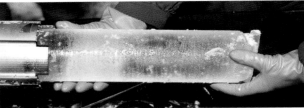

ICE CORE (*far left*) contains bubbles that reveal the composition of the ancient atmosphere. The gas bubbles in the micrograph of a thin cut (*left*) are dark in color and one to three millimeters across.

Another explanation that has been suggested involves the gas in a form known as methane hydrates [see "Flammable Ice," by Erwin Suess, Gerhard Bohrmann, Jens Greinert and Erwin Lausch; SCIENTIFIC AMERICAN, November 1999]. These develop at high pressure, such as that found on the ocean floor. An unknown but possibly very large quantity of methane is trapped in this form in ocean sediments. The sudden release of large volumes of methane from these sediments into the atmosphere has been suggested as a possible cause for rapid global warming events in the earth's distant past. Yet recent results from polar ice core studies show that marine methane hydrates were stable at least over the past 40,000 years, indicating that they were not involved in the abrupt increases of atmospheric methane during the last glacial cycle.

We know that terrestrial vegetation is very sensitive to environmental changes, and thus the total amount of vegetation on the planet varies as the climate cools down and warms up during glacial cycles. In light of our findings, such variations should now be seriously considered as a possible cause of declines in methane levels during glacial periods and rises during the interglacials. During the last glacial maximum—around 21,000 years ago—the plant growth of the Amazon forests was only half as extensive as today, and tropical vegetation might thus have released much less methane. Since that time, global surface temperature and carbon dioxide concentrations have risen, leading to enhanced plant growth and, we would expect, to more and more methane released from vegetation.

Similar climate scenarios may have occurred during other periods of the earth's history, particularly at mass extinction events, such as the Permian-Triassic boundary (250 million years ago) and the Triassic-Jurassic boundary (200 million years ago). Extremely high atmospheric carbon dioxide concentrations as well as rising temperatures could have resulted in a dramatic increase in vegetation biomass. Such global warming periods could have been accompanied by a massive release of methane from vegetation and by more heating. Though speculative, the assumption that emissions may have been as much as 10 times higher than at present is not totally unreasonable. If this is so, methane emissions from vegetation, in addition to emissions of the gas from wetlands and perhaps from the seafloor, could be envisaged as a driving force in historic climate change.

Media Misinterpretations

WHEN YOU SEE A REPORT on your scientific work on the BBC World News immediately following news about bird flu and the situation in Iraq, on the very day your work has first been published, you realize that you have found something with great societal relevance. This realization was reinforced the next day as our research appeared in newspapers around the world, often in front-page headlines.

Unfortunately, extensive media coverage can lead to exaggerations, and in our case it resulted in the misinterpretation of our results. In particular, many reports claimed that plants may be responsible for global warming; in one instance, we saw the headline "Global Warming—Blame the Forests" on the front page of a reputable newspaper.

When you then receive many e-mails and phone calls from individuals asking whether they should cut down all the trees in their garden to fight global warming, you realize that something has gone badly wrong in the communication to the public. We felt compelled to issue another press release to address the misinterpretations.

In our second press release we emphasized that if our finding is true, plants have been emitting methane into the atmosphere for hundreds of millions of years. Those emissions have contributed to the natural greenhouse effect, without which life as we know it would not be possible. Plants are not responsible, however, for the dramatic increase in methane concentrations since the start of industrialization. This surge was brought about by human activities.

Our discovery also led to intense speculation that methane emissions by plants could diminish or even outweigh the carbon storage effect of reforestation programs. If that were correct, it would have important implications for countries attempting to implement the Kyoto Protocol to minimize global carbon emissions, because, under the protocol, tree-planting programs can be used in national carbon dioxide mitigation strategies. But our calculations show that the climatic benefits gained by establishing new forests to absorb carbon dioxide would far exceed the relatively small negative effect of adding more methane to the atmosphere (which may reduce the overall carbon uptake of the trees by 4 percent at most). The potential for reducing global warming by planting trees is most definitely positive.

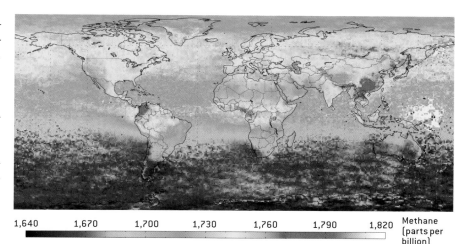

SATELLITE IMAGES of the earth's atmosphere provided support for the authors' controversial finding. In 2005 environmental physicists observed clouds of methane over tropical forests. Although the standard model of methane production cannot explain this observation, the authors' discovery made sense of the curious clouds: the abundant green vegetation of the tropics was emitting the methane.

In the heat of this debate, people forgot a crucial fact: plants are the green lung of our planet—they provide the oxygen that makes life as we know it possible. They perform many other beneficial tasks as well. As just two crucial examples, they provide a natural environment that fosters biodiversity, and they control the tropical water cycle. The problem is not the plants; it is the global large-scale burning of fossil fuels.

A more legitimate concern is whether the methane produced by vegetation can have an impact on climate in the near future. Although plants are not responsible for the massive increase of methane in the atmosphere since preindustrial times, they do tend to grow faster. As we can expect methane emissions from vegetation to increase with temperature, this would lead to even more warming. This vicious cycle would be a natural phenomenon except for its speed, which is accelerated mainly by anthropogenic activities such as burning fossil fuels. The large plant feedback to global climate change that most likely happened in the past, however, is probably unlikely today because so many forests have been cut down.

Although it is too early to say exactly how our revelation might influence predictions for climate change in the more distant future, it is clear that all new assessments should consider emissions of methane by plants. **SA**

MORE TO EXPLORE

The Changing Atmosphere. Thomas E. Graedel and Paul J. Crutzen in *Scientific American*, Vol. 261, No. 3, pages 58–68; September 1989.

Climate Change 2001: The Scientific Basis. Edited by J. T. Houghton, Y. Ding, D. J. Griggs, M. Noguer, P. J. van der Linden, X. Dai, K. Maskell and C. A. Johnson. Cambridge University Press, 2001. Available online at www.ipcc.ch/

Methane Emissions from Terrestrial Plants under Aerobic Conditions. Frank Keppler, John T. G. Hamilton, Marc Brass and Thomas Röckmann in *Nature*, Vol. 439, pages 187–191; January 12, 2006.

Methane Finding Baffles Scientists. Quirin Schiermeier. Ibid., page 128.

Questions for Review

TEST YOUR COMPREHENSION

1. What is the best way for scientists to find out how the levels of atmospheric methane changed over the past 1 million years?
   a. By measuring the width of tree rings and knowing that growth (wider rings) is promoted by warm, wet climates
   b. By studying the yields of domestic and wild rice, knowing that yields increase in warm, wet climates
   c. By studying the thickness of annual layers of ice in the ice sheets of Greenland and Antarctica
   d. By examining the ratio of deuterium (a hydrogen isotope) and standard hydrogen in strata obtained from peat bog cores
   e. By measuring levels of methane present in air bubbles trapped in ice cores of known age

2. Before the discovery that plants produce methane, it was difficult to explain the observation that methane levels
   a. remained constant in the face of climate change.
   b. fluctuated in ways that were independent of periods of relative warm and cold.
   c. fluctuated in step with ice ages, so that there were relatively low levels of methane in cold periods and relatively high levels in warm periods.
   d. fell when there was abundant vegetation and rose when vegetation was sparse.
   e. rose nearly continuously starting 100,000 years ago to reach the high levels seen today.

3. A significant concern about atmospheric methane is that, relative to carbon dioxide,
   a. methane is more than 20 times as effective in trapping heat.
   b. methane destroys ozone more effectively.
   c. methane is more toxic to humans and animals.
   d. methane is more combustible.
   e. methane provides greater support to plant growth.

4. For scientists to detect methane produced by plants, it was necessary to
   a. develop more sophisticated methane-detecting instrumentation.
   b. remove all methane from the air in which plants were incubated.
   c. treat plants with chemicals that physically link newly produced methane to the plant.
   d. dry plant tissue to remove the water that interferes with methane production.
   e. maintain moisture in plant tissues to prevent methane desiccation.

5. Most methane production by plants occurs
   a. in the tropics.
   b. in desert regions.
   c. in temperate zones.
   d. in the polar regions.
   e. over the open oceans.

6. Independent support for the idea that plants produce methane came from
   a. satellite measurements.
   b. measures made by oceanographic vessels sailing over open water.
   c. ecologists working in rainforest tree canopies.
   d. foresters measuring methane output in forests of the Pacific Northwest and Alaskan coasts.
   e. laboratory studies showing that decaying plants emit chloromethane.

7. A source of trapped methane (unknown in size, but possibly very large) is
   a. permanent ice sheets in polar regions.
   b. decaying wood, particularly in temperate regions.
   c. methane hydrates in ocean sediments.
   d. the Great Lakes.
   e. carbon dioxide reserves that can be converted to methane.

8. Plants promote global warming by releasing methane, but they also help prevent global warming by
   a. absorbing carbon dioxide.
   b. absorbing oxygen.
   c. releasing oxygen.
   d. contributing to biomass.
   e. respiration.

9. On balance, forests
   a. contribute to global warming.
   b. neither contribute to nor slow global warming.
   c. slow global warming.

10. If all non-crop plants were destroyed in an effort to reduce global warming by methane emissions, then
    a. global oxygen levels would decline sharply.
    b. global carbon dioxide levels would decline sharply.
    c. methane emissions from ruminant animals and bacteria would rise to maintain equilibrium levels of methane.
    d. methane emissions from methane hydrate deposits would increase to maintain equilibrium levels of methane.
    e. a new ice age would be triggered within decades of plant elimination.